Mogens True Wegener

NON-STANDARD RELATIVITY

A PHILOSOPHER'S HANDBOOK OF HERESIES IN PHYSICS

New Concise Edition, Revised 2021.06.09

Entangled Galaxies, Arp 87
"Hubble Heritage"
Courtesy of NASA & STSci

BOD

Mogens True Wegener

NON-STANDARD RELATIVITY
A PHILOSOPHER'S HANDBOOK OF HERESIES IN PHYSICS

New Concise Edition, Revised 2021.06.09

ISBN

9 788 743 031 420

written in **EXP** by the author
(www.expswp.com)
and printed and published by

Books on Demand
Copenhagen Danmark
www.bod.dk/shop
Price: 111 Dkr

BOD

FIGURE 1.

The dispersion of galaxies
according to Hubble's law
(velocity $\propto$ distance)

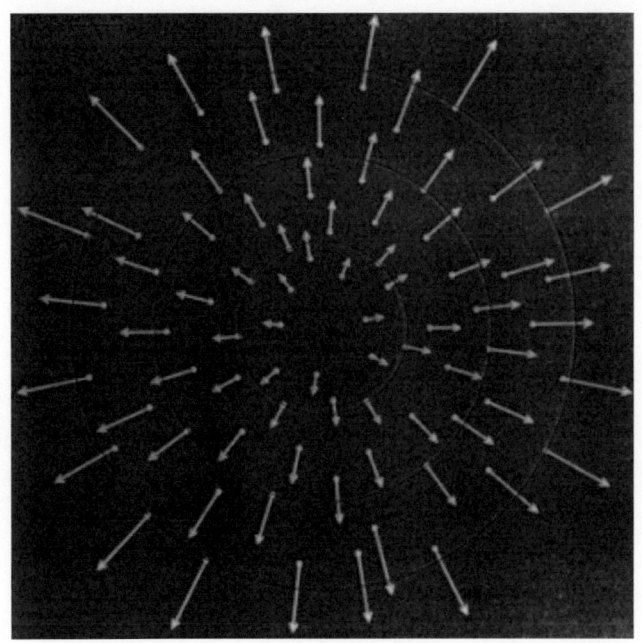

=//=

FIGURE 2.

The Milne "big bang" model,
a pseudo-sphere expanding with the speed of light.
Remove the arrows, and you find the
static pseudo-sphere of my new "steady-state" model.

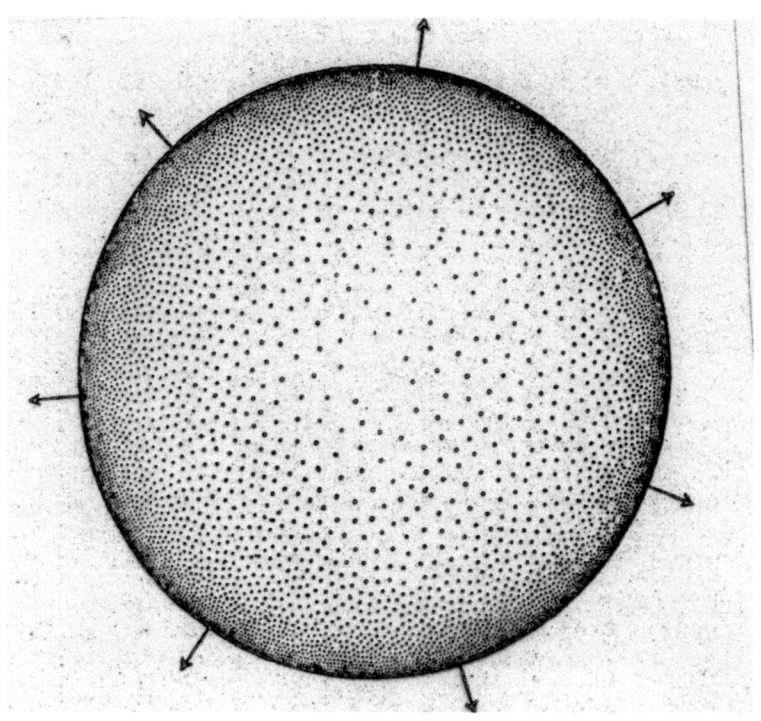

=//=

FIGURE 3.

M.C. Escher: 'Circle Limit 4'
a most wonderful illustration of the
shrinking of galaxies with distance
in a flat space of finite radius.

=//=

=//=

Heard melodies are sweet,
but those unheard are sweeter;
therefore, ye soft pipes, play on;
not to the sensual ear, but, more endear'd,
pipe to the spirit ditties of no tone: ...

'Beauty is Truth, Truth Beauty' - that is all
ye know on earth, and all ye need to know.

John Keats: 'Ode on a Grecian Urn'
(find the urn on the next page)

=//=

CONTENTS

\ /

"The essence of scientific freedom is the right to come to conclusions which differ from those of the majority."   E.A. Milne: Modern Cosmology .., 1952.

"In fact there is no experimental evidence at all for the theory (i.e. Special Relativity); all that appears to support it does so through a circular argument."
H. Dingle, introduction to Bergson, 1965.

"If science is not to degenerate into a medley of ad hoc hypotheses, it must become philosophical and must enter upon a thorough criticism of its own foundations."
A.N. Whitehead: Science & the Modern World, 1925.

"We come thus finally to what is perhaps the most destructive aspect of any physical theory that is 'too succesful' in the social or political sense - it destroys human freedom, and particularly the most precious one, the freedom to think. Forbidden thoughts, censored in their cradle - in this instance typified by the concept of distant simultaneity - always slip in by the back way, to the total confounding of rational thought processes .."
T.E. Phipps, jr.: Heretical Verities .., 1986.

"Imagine two clocks ... which are permanently keeping a perfect agreement. This may happen in three different ways. The first way is to presuppose a natural, or causal, influence (this is the way of the vulgar philosophy) ... The second way to make two clocks agree is to let them be controlled by a skilled craftsman who permanently adjusts the one to the other (this is the way of the occasionalist philosophy). The third way consists in adjusting their mechanisms so well from the beginning that this alone is sufficient to secure their agreement (this is the way of the pre-established harmony)."   G.W. Leibniz: Eclairciss. du Nouveau Systeme, 1695.

"Leibniz's universe was composed of monads which he regarded as mutually independent but his famous principle of pre-established harmony stipulates that the states of all monads at every instant correspond with each other. Leibniz illustrated this principle by the simile of two clocks that have been so perfectly constructed that they keep perfect time with each other without either mutual influence or external assistance. Consequently, in so far as the temporal aspect of the universe is concerned, Leibniz's principle of harmony is equivalent to the postulate of a single universal time. We must therefore discard this principle if we are to reconcile Leibniz's way of regarding time with Einstein's theory of relativity."
G.J. Whitrow: What is Time? London 1972.

It is a tacit assumption of all physics that atoms of the same type, if exposed to the same conditions, oscillate at the same natural rate. Whenever we make use of atoms as "Zeitgebers" in atomic clocks we exploit their Leibnizian Harmony. As hinted at by Leibniz it is non-sensical, indeed vulgar, to ask for a causal explanation of this fundamental fact. Moreover, the standard metric of modern cosmology makes use of a temporal parameter which serves as a cosmic time. So why not simply accept that time is universal, and simultaneity absolute, even though this would necessitate a radical re-interpretation of relativity theory?
Mogens True Wegener

Nature's Code: "We only comprehend the part by connecting it instantly to the whole."
Peter Rowlands

PREFACE

The author of the present book has spent many years on a deep study of relativity theory and its physical interpretation, applying a range of skills - philosophic, scientific, mathematical, historical - that are seldom combined in a single individual. He believes, rightly in my opinion, that it is time to remove the mythology which has always surrounded relativity theory and to establish its real scientific basis on a logically coherent foundation. One of the key aspects of his analysis is a new understanding of time and its place in cosmology. He has been influenced in this by the British tradition of relativistic cosmology, established by E.A. Milne, and developed by his associates A.G. Walker (of Robertson-Walker metric fame) and G.J. Whitrow, and has made a strong argument for its continued relevance.

He believes that, as a philosopher, it is his obligation to search for a way of reconciling the diverging traditions represented by conflicting mathematical and physical interpretations. Only in this way can we arrive at an understanding of the deeper questions concerning the nature of space and time and their relation to cosmology that still remain unanswered at the end of the great twentieth-century physics project. One of Wegener's most significant findings is that, using a version of Milne's kinematic relativity, it is possible to create a world model which can be tested against experience, and which makes sense of both 'big bang' and 'steady state' aspects of cosmology. He has also shown that this cannot be done without an understanding of time which is deeper than any available in the current literature.

Although he has built his argument on a clear philosophical foundation, he has also shown that it can be presented in a rigorously mathematical and logical form. Of course, while some questions can be answered, others still remain, and one of the most fascinating parts of the book comes in chapter 9, where the extent of our knowledge of some of the most fundamental questions ever asked by philosophers, or by scientists - *What is Truth? Is the World Real? Is the World Just One? Is Nature Governed by Laws? Are Occurrences Predestined?* etc. - is put to the test of the methodology developed in the earlier chapters. Current physics and cosmology have left us with a confusion of empirically-derived models which do not yet combine into a coherent structure and which are most often *ad hoc*. Wegener is surely right in believing that the only way to tackle this problem is to go back to the foundations, and he has used skills acquired in many different disciplines to show us that a coherent solution is now within our grasp.

I thoroughly enjoyed reading the book.

Peter Rowlands
Physics Department
University of Liverpool

BRIEF PREAMBLE

In my opinion he (Einstein) would be one of the greatest theoretical physicists of all times even if he had not written a single line on relativity ...   Max Born [1950]

With this quotation I believe to have paid my due homage to the great man. Nevertheless, it is the firm conviction of the present writer that the significance of Einstein's contribution to relativity theory has been grossly exaggerated, partly due to a media hype bursting forth shortly after *WW1*, partly due to the need of the contemporary lay public to admire a non-martial hero. Since then his fame has reached mythological heights. But science is incompatible with myth.

As regards Special Relativity, it has been argued convincingly by Keswani [1964/65] that Poincaré had the entire theory, and that he had it all before Einstein. Against Grünbaum [1960], who has criticized Poincaré for lacking the candour and courage to face the full revolutionary import of the theory, it can be objected that the hesitation of Poincaré to abandon the Newtonian presuppositions of classical mechanics only shows his more mature judgment of the situation.

With respect to General Relativity, it was stressed already by Whittaker [1947] that it is "unwise to accept a theory hastily on the grounds of agreement between its predictions and the results of observations in a limited number of instances". In view of the more recent studies of the theory by Rowlands [1994, 2007] demonstrating in depth that "the revolutionary insights" of General Relativity not only *can* be reproduced on non-relativistic, almost Newtonian, premisses, but that they *were* in fact anticipated long before 1916, Whittaker's remark is most appropriate. In ch.5, §9, I have demonstrated how the (excess) advance of a planetary periastron, as well as the (excess) bending of light rays, are deducible from a few exceedingly simple assumptions.

So there is no binding reason to accept the widespread view that time is not universal, but only the fourth dimension of spacetime, that gravity is nothing but an effect of spatial curvature, and that inertia is nothing but an effect of gravity. Neither is there any reason to believe in the existence of "timewarps" or "wormholes", except those in the brains of brain-washed physicists. Many of the ideas derived from Einsteinian relativity are nothing but phantasy and mythology; but not all Einsteinians were seized by dark orthodoxy or hampered by blind dogmatism.

My late mentor André Mercier made three assertions which I take to be very important: 1) "Gravitation *is* Time" [1975]. 2) "There is no such thing as 'real space'" [1979]. 3) "Spacetime should be reconstructed as Timespace or Supertime" [2000]. Indeed, nothing of this is orthodox. I can only hope that he would not have entirely denounced this handbook of heresies.

=//=

SHORT INTRODUCTION

The present *Handbook of Heresies* consists of three of my contributions to the biennial conferences on the *Physical Interpretations of Relativity Theory (PIRT)*, London 1988-2002, supplemented with two papers delivered at the *PIRT*-conferences in Budapest 2005 & 2007, plus two more papers, one published in the *Foundations of Physics 34*, 2004, the other presented at the *1st International Poincaré Conference*, Nancy 1994, and published by *ACERHP* 1996.

Chapter one, entitled *Relativities at Variance,* is a philosopher's report from the first *PIRT*-conferences 1988-1998. Criticizing two different attitudes, considered as ideal types, viz., naïve formalism and naïve realism, or materialism, I discuss three different ways of interpreting the theory of special relativity or *la théorie de la relativité restreinte*, as Poincaré baptized it: 1) *the spacetime approach*, 2) *the substratum approach*, and 3) *the constructive approach*.

Chapter two, entitled *The Idea of a Cosmic Time*, and dedicated to Franco Selleri, is my attempt to debunk "Einstein's unfinished revolution" (Davies) which seeks to eliminate time, in line with his program for science, viz., "to reduce everything in physics to spacelike concepts". An earlier version of the paper was published in *Foundations of Physics 34*, pp.1777-99, 2004.

Chapter three, entitled *Milne's Kinematic Relativity*, was originally printed with the title *Ideas of Cosmology, A Philospher's Synthesis*, as my own contribution to the collective volume *Recent Advances in Relativity Theory I*, Duffy & Wegener eds., printed by Hadronic Press 2000. It contains the result of my efforts to vindicate the *British Tradition of Relativistic Cosmology* as represented by the names of E.A. Milne and his assistants A.G. Walker and G.J. Whitrow.

The fourth paper, *Some Cosmological Models, their Time Scales and Space Metrics*, was presented at the 2002 *PIRT*-conference in London. Most of the cosmological stuff is standard, except that my new model of a universe based on Continued Creation is briefly sketched.

The fifth paper, entitled *Big Bang versus Steady State*, was presented in Budapest 2007. It elaborates on the model hinted at in chapters 3-4, from the perspective of kinematic relativity, demonstrating that the idea of a "big bang" is not incompatible with that of a "steady state".

The sixth paper, entitled *New Axioms for Cosmology*, was presented in Budapest 2009. It contains my attempt to reconstruct spatial geometry from a set of purely temporal axioms, in accordance with the program of André Mercier, viz., *to reconstruct spacetime as timespace*

The seventh paper, bearing the title *Constructivism in Science*, was first published 1996. This paper outlines the philosophy of science advanced by Poincaré, with inspiration from Kant, followed by a short sketch of some similar thoughts to be found in Eddington and Milne.

Article eight, entitled *Fundamental Queries*, contains my recent reflections on some basic problems of a more philosophical, or even "metaphysical", character. Philosophically minded readers might profitably begin with this survey of "deep questions" in modern cosmology.

In general, I recommend my readers to ignore most of the mathematics at a first reading. Finally, I apologize for repetitions and overlappings between the various chapters of the book. When read with friendly eyes, its contents might be described as: *thema con variazione.*

=//=

Non-Standard Relativity

CHAPTER 1

RELATIVITIES AT VARIANCE
CONFLICTING IDEAS OF RELATIVITY THEORY

*A Philosopher's report from
the PIRT-conferences 1988-1998,
Imperial College, London.
Revised 2015 & 2021*

As one of the rather few philosophers attending the biennial conferences on the *Physical Interpretations of Relativity Theory* right from the beginning in 1988 until 1998, ten years later, I would like to take this opportunity to assess the import of the various contributions offered by these conferences from a philosophical point of view, in order to give some clues as regards the perspectives of future progress in this field of physics. Throughout the past decade, all our conferences have been kindly sponsored by the *British Society for the Philosophy of Science*. For that reason it may not seem imposterous for a philosopher to assume this task.

Dr Duffy, the glorious initiator and excellent secretary of these conferences, has also chosen their title and written the text of their programme, from which I quote: "Contributors should note that the starting point of the conference programme is the acceptance of the accuracy and excellence of Relativity Theory which provides the framework for the discussion. The questions raised are directed towards examining the philosophical, methodological and historical aspects of the various interpretations of the formal structure, and the implications which these several interpretations have for physical theories". No one, however, could rightly accuse Dr Duffy for having exerted the harsh strictures of blind orthodoxy and I, for one, am gratified that the actual course of these conferences has not shown any resemblance to what one might have feared from a very rigorous implementation of the passage just quoted. In fact, our little society has rather witnessed "the flowering of a thousand flowers", in contrast to those much larger societies which have been marred by the crippling influence of dogmatism.

But not everything is well in our little garden, part of the great field of natural science. Relativity is a topic in mathematical physics. I remember from a conversation with a collegue from the mathematics department at our university how one of his casual remarks struck me with surprise. Mathematical physics, he told me, is an altogether different subject depending on whether it is practised by a mathematician or by a physicist. In fact, mathematicians and physicists constitute two different camps within the field of mathematical physics: each camp having developed its own scientific journals and organizations, their mutual communication and intercourse is next to nil. Metaphorically, the two camps are separated by a huge distance covering a vast desert of wasteland too hostile to the passage of visitors.

Mogens True Wegener

This division into two camps has its counterpart in a cleavage that threatens to split our own society into sections that do not communicate. So their adherents sometimes behave like sectarians. One could say that such conflict of interpretations is the price we have to pay for our freedom of thought. In a way I believe that this is true; but it is a philosopher's task to work for unification wherever unity is possible, and I do not see any reason for giving up in advance. The division, alluded to by my collegue, between mathematicians and physicists, I see repeated in our own forum as the cleavage between two divergent attitudes to issues of relativity theory.

These attitudes, exposed as *ideal types*, I shall henceforth characterize as **naïve formalism** and **naïve realism**, respectively. So I put up two ideal types as virtual targets in order to shoot them down by intellectual criticism. By treating the two attitudes as radical extremes, I hope that I can attack them freely without running the risk of hurting anybody personally. Luckily, these ideal types rarely manifest themselves historically in a pure form; but the divergence of trends is clear, although their statistics is a far cry from equilibrium. In fact we find a marked tendency towards the preponderance of realists among the participants in our meetings which may be due to the presence of the predicate 'physical' in the title given to them.

Attitudes towards science are not themselves scientific. Neither can arguments advanced in support of such attitudes be called scientific, although their proponents attempt to underpin them by reference to what they believe is well established science. What unifies the two attitudes just mentioned as opposite poles within a single field of tension is *their common quest for an ultimate explanation*, meaning an explanation of our sensible experiences which has its foundation in "the hidden depths" of *Nature*. Therefore they remain on a par as regards their philosophical urge towards **ontology**, or **metaphysics**.

What differentiates them is merely the sort of explanation they suggest. *a)* In the case of *naïve formalism* the explanation consists of a reference to the immanence in *Nature* of eternal forms, mathematical or geometrical. So naïve formalism is just *pure idealism*, the most vulgar kind of Platonism, or Pythagoreanism. *b)* In the case of *naïve realism* the explanation involves the claim that *Nature*, in reality, consists of matter-in-motion, of non-sensible fields, or of a substance, or substratum, termed the aether. Thus understood, naïve realism may stem from the Ionic search for a material principle, from the atomism of Democritus, from the aether-theory of Descartes, or from some other source. *Arché* being the Greek word for a principle of origin, we may describe naïve realism as *archaic*. Since formalists often take over the *jargon* of realism, by their speaking of "curved spacetime" or "the structure of pure vacuum" as ultimate realities, the issues sometimes get somewhat blurred. Hence, instead of speaking of realism in contrast to formalism, I might have spoken of *materialism*, or *substantialism*. In that case I would not have needed the predicate 'naïve' since, today, all traditional kinds of materialism, or substantialism, are most certainly naïve. However, what I here brand as *naïveté* is purely philosophical, viz., a peculiar kind of intellectual immaturity that, nevertheless, remains compatible with the highest degree of scientific sophistication and ingenuity.

The Urge for Metaphysics as Ontology
Naïve Formalism (vulgar Platonism/Pythagoreanism)
seeks to unveil the "natural geometry" immanent in the depths of Nature
Naïve Realism / Materialism (the Ionians, Democritus, or Descartes)
seeks to disclose the "underlying reality", the "hidden substance", of Nature

Non-Standard Relativity

How do these different positions manifest themselves in the context of our conferences? This can best be illustrated by reference to a now famous little book entitled *The Logic of Special Relativity*, Cambridge 1967, due to one of our most prominent collegues and friends, the late Simon Prokhovnik. In his monograph professor Prokhovnik discussed three seemingly very different interpretations of ***SR***, based on: 1) *the logic of spacetime* (standard), 2) *the logic of relative motion* (kinematic relativity), 3) *the logic of absolute motion* (substratum theories). For my part, I regard the kinematic (constructive) approach as a proper mean of the other two:

    a) The SpaceTime Approach - seeks the inherent structure of space
    b) The Substratum Approach - hunts the ultimate substance, or frame
    c) The Constructive Approach - invents world-models from first principles

It is characteristic for the presentations and discussions delivered at these conferences that a majority of the participants are adherents of *realism* and the *aether-theoretical* approach and that a considerable part subscribe to *formalism* and the *spacetime geometrical* approach, while almost nobody except myself pays any interest to the approach of kinematic relativity. This I find regrettable for at least two rather important reasons.

First, a dogmatic accentuation of one of the opposite views to the exclusion of the other would immediately transpose us to that philosophical brand of naïveté which I rightly denounce. I fully acknowledge the value of the formalistic approach of *spacetime geometry as a technical instrument of relativistic physics*. Only I side with my mentor André Mercier who in his 1994 PIRT-lecture on "The Reconstruction of Spacetime as Timespace" insisted that time should be considered more important than space, so that 'spacetime' should be reconstructed as 'timespace' and that it is illegitimate to hypostasize geometry into a *structure* which is immanent in Nature. Likewise, I fully recognize the value of the realistic approach of *aether-theory as a heuristic device to be utilized* in order to further the fruitful development of relativity theory; but it is also illegitimate to hypostasize the aether into a *substance* underlying the existence of Nature.

Since the relevant questions of modern science are questions of structure rather than questions of substance, it is easier to unmask the mistake of realism than the fault of formalism. However, viewed as extremes, these positions are glaring transgressions of the limits of science. *As argued convincingly by Kant, all sober science should abstain from statements of ontology.* Thus, if one of the approaches is considered in spendid isolation from the other, and its content is elevated to the exclusive status of ultimate reality, or essence of nature, then I must object. Philosophically, this inborn ontological urge, this drift towards a deep metaphysics, in short: this *deep naïveté*, is not only misleading in the sense that it takes us out in a swamp in order to leave us there with a mess of inconsistencies and unsolved problems: it is also dangerous in the more serious sense that it tempts us to seek our refuge in scientific dogma.

Second, I suspect that the two conflicting attitudes, that of formalism and that of realism, secretly uphold a sort of unholy alliance in the sense that they conspire to ascribe a conceptual primacy to space rather than to time. In this they just follow the lead of Einstein who explicitly stated that his scientific program was to reduce everything in physics to space-like concepts; but in that respect, at least, the tradition from Einstein is obsolete. As a historian of ideas I can affirm with confidence that, just as it was the unique feat of the renaissance to discover space - mind the introduction of perspective in the arts - so it is up to our own century to discover time - cf. the telling title: *From Being to Becoming*, given by Prigogine to a reknown book of his.

Mogens True Wegener

The great minds behind the classical revolution of natural science, Galileo and Kepler, were unanimous in their assent to these words: *ubi extensio, ibi materia, ibi geometria* - where there is extension there is also matter and geometry. The same stand was taken by Descartes who was unable to make his analytic geometry relevant to physics without calling upon an aether theory. This is also the reason why I can imagine a secret conspiracy between realists and formalists, between aether theorists and spacetime geometricians: both parties consider time to be a mere illusion, both parties want to exclude it from serious consideration and analysis.

Am I exaggerating? I very much hope so! For **Time**, which is *the inmost gist of creation, freedom, life, can't be excluded!* To exclude time would be to divest the world of its dynamics. Can we imagine a timeless world, a world without change? Such a universe couldn't be real! The Greek philosopher Parmenides (*floruit* 500 BC) seduced himself into believing that he was able to imagine a block-universe wholly devoid of time and motion; but I am convinced that it was an illusion made possible by his incorporation in a changing world where he could think and reason and thus partake in temporal processes. His contemporary, Herakleitos, had a much more realistic picture of the world, imagining it to consist of a steady flow of fire, sometimes bursting up and sometimes fading away, but always ruled by divine decree, or law (*lógos*). Before him, Anaximandros (*floruit* 560 BC) took interest in time when describing the principle (*arché*) and element (*stoicheion*) of things as the infinite, or indefinite (*tò ápeiron*) "that gave origin to the heavens and the worlds within them" and which is still "the source of becoming as well as destruction"; and, as he further added: "all things change with necessity" for they "pay penalty and reprisal to each other for their crimes according to the judgment of Time".

But what of space? Cannot the void be "real"? In the *Timaeus*, his cosmology, Plato spoke of three "things" (*ónta*): 1) *Pure Being*, which as *timeless form* is the object of *reason*, 2) *Pure Becoming*, which as *temporal events* is the object of *sensation*, and 3) *Pure Void*, which as the *receptacle* of becoming-simulating-being is *dreamlike*, neither an object of pure reason, nor one of pure sensation, hardly one of belief. Has our enigma ever found a finer expression?

> **Pure Being** (*tò ón*) = **eternal forms**: *ideas, geometry - objects of reason*
> **Pure Change** (*génesis*) = **temporal events**: *phenomena - objects of sensation*
> **Pure Void** (*chóra*) = **the uterus of creation**: *in between - dreamlike, object of neither*

According to Plato, these are timeless conditions for the temporal existence of World (*kósmos*) which was unified with Time (*chrónos*) by decree of the Divine Master Craftsman (*demiourge*) right from the dawn of creation. In the same passage he wrote of the Void (*chóra*): "Third is Space, which is everlasting, not admitting destruction, providing a situation for all things that come into being, but itself apprehended without the senses by a sort of bastard reasoning *(sic!)* and hardly an object of belief. This .. we look upon as in a dream, saying that anything that is must needs be in some place and occupy some room." (Cornford's translation)

Now, what is the verdict of modern relativistic physics? Let us, for instance, consult the book: *On General Relativity*, by Mercier, Treder & Yourgrau (Berlin 1979). Herein the concept of space is discussed, and the authors argue from a plurality of possible spaces to the conclusion that space is not real: "(A) space must be constructed from a suitable axiomatics. Axioms are not evident truths, they are implicit definitions. Therefore, none of these spaces is 'real space'. *There is no such thing as real space"* (my italics, MTW), This stance, which I entirely share, supports the view of Poincaré: *space in itself is devoid of structure*. But is space an illusion?

Not quite! Rather it is "pure possibility" as proposed by Aristotle, or "well-founded appearance" as suggested by Leibniz, a late pupil of Plato. So, in what do we find its foundation?

In my view: *physical space is timespace, i.e., a modification of time*. We may have time without space: that would be something like a particular relativistic *world-line*, the symbol of existence of a material particle, or of a human observer. But we cannot have space without time; that would, quite literally, be sheer nonsense. Space is *multiplicity unfolded across time* or, in logical terms, spatial extension can be defined as: local exclusion of simultaneous events. Such simultaneity is definable by the absence of causal connectivity, "true causes" operating in time by the communication of information-carrying signals propagated at a certain speed, viz., that of light which, moreover, is supposed to be the universally invariant limit to all motion. But our concept of causality depends on, and is derived from, our concept of laws of nature. Further, the distinction between before and after in the relation of causal connectivity, between the cause and its effect, cannot, in my opinion, be introduced without explicitly or implicitly referring to a prior order, or "arrow", of time. The causal theory of time involves a vicious circle, and to explain it by Reichenbach's method of marks is unconvincing; cf. p.112.

Our definition of simultaneity, and that of spatial extension across time, hinges on signal communication. The only relevant signals consist of electromagnetic radiation; whether they be visible or invisible to a human eye, let's for simplicity's sake agree to speak of "light signals". There is, then, an indisputable interdependence between the speed of light and the definability of simultaneity and that of spatial extension. But how speak of a "speed of light", how suppose light to be "something travelling" in space, even before having defined space? In this forum it is unnecessary to elaborate on the well-known circularity inherent in the attempt to determine the one-way light-speed by timing the arrival of a signal at a distance by means of a clock made synchronous to the master-clock of the emitting observer by another exchange of light-signals. What is operationally feasible is solely the timing of the interval between the advanced and retarded times of a reflected ("radar") light-signal, or of series of such signals.

Hence, as stressed by an impressive number of papers during the past decade including e.g., those of Kroes, Selleri, Sjödin, Sklar, Øhrstrøm, and myself, two of the main conclusions of this forum seems to be: 1) that the only interpretation of Einstein's light principle relevant to physics is the one asserting a universal constancy of the round-trip, or average, light-speed; and: 2) that an indefinite number of non-standard definitions of simultaneity, all involving variable one-way light-speeds, and thus at variance with Einstein's own convention, are indeed possible. Thus the issue of simultaneity remains a stumbling block to any "realistic" view of space.

For that reason I accept the suggestion of Viv Pope [1996] that we consider the phrases "light-speed" and "velocity of light" to be dubious metaphors which, for philosophical purposes at least, should be replaced by a much more precise linguistic usage that simply refers to the universally invariant proportionality between temporal and spatial standard intervals, or units. Likewise I accept his proposal that we regard "photons" as *binary quantum relations* which are *not in themselves spatial*, but which may be useful to the purpose of introducing space, since light, albeit *retarded relative to frame time*, may yet be *instantaneous relative to proper time*. However, when Pope rejects what he calls a "God's Eye View" of cosmic symmetry, accusing it of having mislead him to think that there was a clock paradox, I simply don't understand him. According to Pope, it is not merely logically impossible to think of three observers in relative motion as being in a situation of perfect symmetry, it is also (it seems) a kind of blasphemy ...

Mogens True Wegener

I quote from his [1994]: "What (standard, MTW) relativity was telling me was that so long as we drop that presumptuous, socially conditioned belief that the way we see things in our mind's eye is the way we imagine 'God sees it', then there is no paradox whatever in time being different for different observers. All we have to do is to settle for the fact that, in reality, one can only describe what happens in the time by which one sees it happen. In that relative time, objects which are in motion relatively to oneself age at different rates ... I saw very clearly that if observational distance is observational time in the ratio of units c, then the times ticked by clocks that change their observational distance as they tick observational time must, logically, be stretched-out, or 'dilated' relatively to the observer, in the *geometrical* way (**SR**) describes." This amounts to a complete denunciation of the cosmological approach of kinematic relativity. But, surely, what has mislead Pope is the approach of spacetime geometry.

Since my intention is to invite you to adopt precisely that "God's eye view" which Pope rejects, I shall make my stance clear: What I invite you to do is not blasphemy, but cosmology! *The limitation of the spacetime geometrical approach is precisely that it is local, not universal.* As I have just demonstrated, what the spacetime approach and the substratum approach have in common is the urge for ontology. But the world as ultimate reality is and remains inscrutable. Never, never shall we know! When the modern picture of the universe as an aetherial machine was developed by Descartes, his scientific imperialism, which left nothing for the humanities to do, was attacked by Giambattista Vico. According to Vico, *verum et factum convertuntur*, what we can know is what we can do, nothing more and nothing less. Now, to create the universe was the feat of God, our Creator. His act was unique and therefore cannot be repeated. What we can hope to understand, as human beings, is therefore history, said Vico, not physics. So the arts, humanities, and culture, are more important than science.

In a way Vico was right, I believe; but I don't share his pessimism with regard to science. In fact, I think that Vico, unintentionally, has presented us with the very key to good science! What we know is what we have done. Hence, to use a metaphor, the task of good science is to retrace the footsteps of God! This was also the clear conviction of Nicolaus Cusanus (1401-64). Following him, I shall insist that *physics* can only realize its deepest aspirations as *cosmology*. The very aim of natural science is to reconstruct the universe and, in order to fulfil its purpose, it must be based on simple principles and clear definitions. [MTW, 1994 & 2000].

Let us start again from scratch. Exact science attempts to discover "the laws of nature". When found, such laws are represented as *invariant relationships* in the stream of experience. Only relationships *confirmed by repeated observation and experiment* can pass as natural laws. Science is a human business which refers to observers: the laws of nature must be valid to all possible observers; hence the importance of invariance, and hence that of the human observer. But couldn't we abstract from observers? Isn't it sufficient to refer to 'events' and to 'frames'? One of my main points is that the concept of *event* is all right, while that of *frame* is not.

This is reflected in the difference between Einstein and Poincaré regarding their various formulation of **the relativity principle**: whereas Poincaré spoke of the **invariance** of the laws of nature with regard to the transformation of coordinates, i.e., the communication of observed data, **between observers**, Einstein instead spoke of the invariance, meaning the **covariance**, cf. Tom Phipps, of natural laws with respect to the transformation of coordinates **between frames**. For my part, I clearly side with Poincaré and his observers as against Einstein and his frames. My point is that *frames are artificial constructs which are do not pre-exist in nature.*

Non-Standard Relativity

To recognize this we must recur to a remarkable British Tradition within relativity theory, represented by the names of E.A. Milne, A.G. Walker, and G.J. Whitrow. While Milne was the intuitive genius of new ideas, Walker and Whitrow were gifted with the analytical talent needed to exploit his exceptional ideas and bring them into full blossom. According to Milne the foremost condition of describing the universe by a rational world-model is that the structure of a world-model is determined by the existence of a universal class of equivalent particles, or observers, *the substratum*. This kinematic substratum *should not be confused with an aether*. In his work *Cosmologie du XXme siècle* [1965], J. Merleau-Ponty has strikingly compared such "particle-observers", or "observer-particles", with Leibnizian monads, describing the kinematic relativity theory of Milne as *a veritable monadology translated into mathematics*!

Indeed, the difference between the kinematic approach and that of aether-theory seems to be abyssmal, at least to begin with. Later, when we have derived the geometrical structure of the kinematic substratum and deduced its consequences for the timing of light-signals, it may appear natural to speak of the "propagation" of some quasi-entities called "photons" and to ascribe them speed, or velocity, both one-way and two-way, at an instant, and on the average; however, we should be wise enough to remind ourselves that such phrases are just metaphors. Whether, at that stage, we find it appropriate to describe the kinematic substratum as a specific type of aether, or we prefer to abstain from such language, is merely a matter of taste.

But, to begin with, following the lead of Milne and Walker [MTW 1996a], our approach is more in line with a *lógos-theory* than with a *physis-theory*, to use the phraseology of Pope. For this reason, the class of world-models encompassed by our approach will display features making a comparison with current research into the structure of information processes natural. The main issue that will concern us is the nature and properties of the kinematic substratum. According to Milne, the structure of the universe is determined by the kinematic substratum, the members of which constitute a privileged class of equivalent observers; laws of nature are, according to his *cosmological principle* (*CP*), invariant to all members of the substratum.

In agreement with this we distinguish between *fundamental observers* which realize an almost perfect equivalence and thus are members of the substratum, and *accidental observers* which are a far cry from perfect equivalence and for that reason do not belong to the substratum. This distinction puts *the idea of cosmic symmetry* to the fore. Without this idea it is impossible to develop a cosmology. The difference between fundamental and accidental is not absolute, but a matter of degree, so the idea of a class of perfectly equivalent observers is an idealization. Let us reflect a little on the contents of *the principle of relativity* (*RP*). What does it state?

Following Poincaré, it states the invariance of all laws of nature to the exchange of data, or the transformation of coordinates, between equivalent observers. In a way this is a tautology, since. if the laws are not invariant, the observers are not equivalent. His formulation makes it natural to view the *CP* as a strong version of the *RP* [MTW 1994b]. In agreement with this, no law of nature can be excepted from the invariance assumed to hold between fundamental observers, especially not those laws that govern the behaviour of clocks. So there is no clock paradox pertaining to the master-clocks of fundamental observers for the very simple reason that all such clocks agree and, if they don't, these observers are just not fundamental.

Therefore I conclude that a cosmic time is indispensable to rational cosmology.

Mogens True Wegener

It is often said that homogeneous and isotropic world-models allow for a cosmic time. I now invite you to invert this order of reasoning. According to the strong **RP**, *alias* the **CP**, the structure of all rational world-models must be determined by the existence of a universal class of privileged observers that are distinguished by their common participation in a cosmic time. On what conditions can this claim be sustained? The only way to exclude all possible influence from external causes is to ensure the most perfect symmetry. Now this means cosmic isotropy. All directions in the universe must be equally "good". No spatial direction can be privileged. Only on that condition can the universal substratum function as a *compass of inertia* (Weyl). But this can only hold for the fundamental observers. For accidental ones it must be different. Hence, for accidental observers, the cosmic symmetry is broken and anisotropy reigns.

This anisotropy, or asymmetry, according to Milne, is the only reason which can be given for the emergence of forces in the universe; all forces, also gravity, must be due to asymmetry. Now experience has shown clocks to be retarded, and light-rays bent, near gravitating bodies. This seems to indicate causal influence, but can't the relationship be seen the other way round? What if it is the retardation of clocks that, by the "bending" of light-rays, hence also of "space", induces massive bodies in free motion to approach each other spatially? This view, at least, makes it plausible to understand asymmetry as being the cause of gravity. In fact, my mentor André Mercier once made a remarkable statement in a paper, the reading of which encouraged me to make contact with him. What he wrote was: "*Gravitation* **is** *Time*" ...

One last point. In his book and in several papers, cf. e.g. the important quotation in the following appendix, Prokhovnik suggested the possibility of a stretching of "light-in-motion" (mind the metaphor!) due to universal expansion. What Prokhovnik here called "the hypothesis of McCrea" was anticipated with at least one year by Whitrow, who in his book *The Natural Philosophy of Time* [1961/1980] suggested a variation of light speed over cosmic distances. This Whitrow did in the context of a discussion of the relativistic formula for clock-retardation from which he claimed to be able to derive the standard Robertson-Walker metric *(RWM)* of modern cosmology. That issue will be treated more fully in Chapter 4, §3.

=//=

APPENDIX:
Important Quotations from
Duffy & Wegener, 2000: *Recent Advances in Relativity Theory*, vol.1.

B. Tonkinson: 'Clocks don't go slow, Rod's don't contract' [PIRT 1996]
    "The notion that there will be different clock rates or changed lengths in different inertial frames is misleading, clocks do not "go slow" and measuring rods are not "contracted"." - cf. Törnebohm below.

P. Kroes: 'The Status of Time Dilation within *SR*' [PIRT 1988]
    "The only way that the notion of an ether can be made compatible with *SR* is to deny that the ether can be ascribed a definite state of motion .. (But) if the ether cannot be ascribed a state of motion then .. Lorentz's conception of length contraction and his dynamical explanation loose their objective meaning."

P. Øhrstrøm: 'Tense Logic and **STR**' §7, cf. Duffy & Wegener, eds. [2000].
    "It is even possible to solve the problem without introducing a preferred reference frame, or a preferred direction, as it suffices to assume the existence of a set of fundamental particles, i.e., a so-called substratum. All fundamental particles are assumed to move inertially, and each fundamental observer is supposed to have his clock synchronized with that of any other fundamental observer by their original coincidence at $\tau = 0$. It is easy to show that this definition corresponds to a re-synchronization according to the convention: $\tau = \sqrt{t^2 - x^2 - y^2 - z^2}$.
    Prof. H. Törnebohm has discussed this interesting idea in several publications. It is evident that τ-time is invariant under *LT*, and τ-simultaneity can therefore be considered absolute on the condition that the τ-scale can be ascribed an absolute zero. Törnebohm sees the "big bang" as a plausible description of the origin of time $(\tau = 0)$. In his opinion, it is tempting to identify the fundamental particles with (the centers of) galaxies. There are two problems associated with this solution.
    The first problem is that (it) presupposes $x < ct$. (Hence) clocks can only be synchronized within a uniformly expanding universe ... (A consequence of Törnebohm's solution is that) the velocity of light varies over space as well as in time. (Hence), at $\tau = 0$ signals have infinite velocity. I (believe) that (Törnebohm's) alternative version of *SR* is empirically equivalent to the ordinary version. I think that this equivalence is very important from a philosophical point of view. In the first place it demonstrates the possibility of an absolute simultaneity that is consistent with all empirical consequences of *SR*. Secondly, it shows that the "light-age argument" needs not be valid, i.e., a light-signal that has travelled over a distance measured to be n "light-years" was not necessarily emitted n years ago ...
    Another problem related to the above solution is that the transformation of space-coordinates is non-linear: $\tau' = \tau$. $x' = \gamma(v) \{x - v\sqrt{\tau^2 + x^2 + y^2 + z^2}\}$. $y' = y$. $z' = z$ ... (This) non-linearity leads to the result that "force-free" particles will in general be accelerated ... The difference between the orthodox version of *SR* and (Törnebohm's) version can easily be explained. Suppose that a light signal is sent from a clock located at $(x, y, z) = (0, 0, 0)$, at the time t_1 to a clock located at (x, y, z), where it is immediately reflected at the clock reading τ_2 to be received at $(0, 0, 0)$ at the τ_3. If the two-way velocity of light is unity, then: $r = \sqrt{x^2 + y^2 + z^2} = (\tau_3 - \tau_1)/2$, $t = (\tau_3 + \tau_1)/2$. It follows that the non-standard time coordinate (is): $\tau = \sqrt{\tau_3 \tau_1}$ .. whence: $t - \tau = t\{1 - \sqrt{1 - r^2/t^2}\}$. "

Mogens True Wegener

S.J. Prokhovnik: 'The Nature and Implications of the Robertson-Walker Metric (*RWM*)' [PIRT 1990]:

"Looking at the *RWM*, we note that it defines a unique cosmological reference frame associated with the set of fundamental observers. The significance of the constant c in this context is unmistakable; it represents the speed of light considered in respect to this particular reference frame; hence, the formulation of the *RWM* clearly implies that the propagation of light takes place relative to the set of fundamental observers, which for this reason defines *a cosmological substratum* ... Both Bondi and Bergmann voice their concern that the existence of a preferred reference frame appears to be in conflict with *SR*. Furthermore, the (idea of a cosmological substratum) implies that light will pass a succession of fundamental observers with the same speed c, irrespective of the expansion of their reference frame as described by the scale factor $R(t)$. *This would mean that light could ultimately reach us from any fundamental particle no matter what its recession velocity, and that it will reach us, albeit redshifted, with the same speed as light emitted from any other origin* (my italics, MTW). In this context, the cosmological Doppler redshift effect can be considered as a direct and intelligible consequence of a light ray's maintenance of its substratum speed in the face of the expansion of the substratum; it is as if the ray is 'stretched' (and thus its energy diluted but not lost) by the expansion, depending on the scale factor $R(t)$. It is seen that this interpretation leaves open whether or not there exist galaxies with recession velocities greater than c, and hence it is also neutral on whether the universe is finite or not.

The idea that there exists a fundamental reference frame, a cosmological substratum, for light propagation - (notice that Prokhovnik, on p.75 of his book: *The Logic of Special Relativity* [1967], spoke of this as: *McCrea's light-hypothesis,* my insertion, MTW) - is by no means inconsistent with physical observations. We know that light is affected by a gravitational field, so we might well expect that its cosmological behavior should be basically determined by the field associated with the overall distribution of matter in the universe. Such a non-arbitrary basis for its propagation would explain why its velocity is independent of its source, an amply-confirmed observation. That the velocity of a light ray should be the same with respect to every fundamental observer in its path is fully consistent with the equivalence of all fundamental observers as required by the Cosmological Principle, and it is also consistent with physical experience that light does not overtake light, signifying that light from a distant source must reach us with the same speed as light from terrestrial (or any other) sources, irrespective of the distance or relative velocity of the source. The light hypothesis enables one to calculate precisely the distance travelled by light relative to its source (treated as a fundamental particle) from the geodesic of the *RWM* for an assumed form of the scale factor $R(t)$. Thus, accepting the *RWM* and its implications explicitly, we are able to interpret astronomical luminosity and redshift observations of distant galaxies more satisfactorily than by evading the notion of a fundamental frame and the associated light hypothesis."

The rest of the passage in **Mercier, Treder & Yourgrau** [1979] runs as follows:

"Yet many a relativist today might be tempted to say: Oh yes, Riemannian space is at least an approximation to real space. But why should not a quantum theorist then say: Hilbert space is such an approximation? The answer might be of course: each quantum system needs another Hilbert space, so this is a fiction, whereas the Universe (the totality of what is) needs something like a Riemannian space, it is even identifiable with it. Then we shall ask: with what Riemannian or other space exactly is it identifiable? give me its metric g_{ik} and all its further properties as final datum, and then everything is determined, is even superdetermined, in it ... It is Spinoza's God, if you will, and we must be pantheists. .. Apart from the uneasiness produced by this eventuality, everyone may have guessed that the kind of revolution which, following the appearance of Einsteinian relativity, has taken place at the beginning of this century, may very well repeat itself .. making *GRG* obsolete and replacing it by some super-theory."

CHAPTER 2

THE IDEA OF A COSMIC TIME
DEBUNKING "THE EINSTEINIAN REVOLUTION"

Revised version (2015,2021) of paper in:
'Foundations of Physics' 34, pp.1777-99, 2004.
Reprinted with kind permission from
Springer Science&Business Media B.V.

=//=

This paper was written in honour of Franco Selleri (1936-2013),
faithful defender of reason in physics, who committed his efforts to
"the liberation of time from the enslavement to space".

=//=

Summary

Pointing to the cleft between the idea of a temporal evolution,
central to modern biology, and the idea of the timelessness of reality,
fundamental to modern physics, the present paper demonstrates that the
standard definition of time at a distance is beset with ambiguities which
might be solved by making a fresh start taking its point of departure
in the idea of an absolute Cosmic Time in accordance with the
British Tradition of Relativistic Cosmology.

=//=

Contents

=//=

Mogens True Wegener

1. INTRODUCTION

In a special issue of *Scientific American* dedicated to ***Time*** [287 no.3 sept.2002] a notable sceptic makes fun of the fact that smart people often believe weird things. The innocent reader may be surprised to learn that this ironical remark, targeting at phenomena such as astrology, clairvoyance, magnetotherapy, and ufology, is also applicable to some allegedly scientific views promoted in that issue, e.g., the opinion reported below:

Scientific American is generally acknowledged to be a serious magazine and Paul Davies, a scientist of high repute, is regarded as one of the more reliable mediators of modern physics. Nevertheless Paul Davies makes himself a spokesman of the opinion that, from the point of view of science, the idea of *temporal flux* is nothing but illusion. He even tries to underpin this view by appealing to the *special theory of relativity*, invoking the creator of that theory as his main witness. Indeed, Einstein made a queer attempt to comfort the widow after his deceased friend Besso by reminding her of the delusive character of all temporal phenomena!

In contrast to physics, which is neither capable of explaining what it means that "time is passing" nor qualified to ascribe a direction to "the arrow of time" (a happy phrase coined by A.S. Eddington), there is a large number of other sciences which not merely presuppose the passage of time, hence also its direction, but which even prosper by describing it. At the same time it is equally clear that these historical disciplines - whether they belong to the arts and humanities, or to the social or the natural sciences - would lack all scientific legitimacy if the metaphor of *time-in-flow* could be shown to be meaningless or indefensible.

Turning our eyes towards a science like biology, it is immediately evident that something as basic as the doctrine of natural evolution must appear completely nonsensical, devoid of any rational meaning, if the very notion of time's passage cannot be accorded any scientific status. That it cannot is argued by J. Barbour in his *The End of Time* [1999], which I take to be the final apotheosis of Einstein's programme: to reduce everything in physics to "spacelike concepts". It is paradoxical and highly problematic that the most recent results of natural science force us to choose between physics and biology. But a similar conflict can be found *within* physics!

Considering the view that modern cosmology began with the general theory of relativity and peaked in the dogma of "big bang" it is not only conspicuous, but sensational, how radically the prevailing cosmological paradigm is countered by the trend to skip time of physical reality. According to the "big bang" hypothesis our universe began some 14 billion years ago in a huge explosion that marked the zero of time. But the arguments against time, in antiquity provided by Parmenides and Zeno, are today derived from .. Einstein's two theories of relativity!

Thus, according to the unanimous verdict of modern cosmology and biology, everything in the world observed or experienced today, including ourselves, is nothing but the prolonged effects of an evolution initiated by the creation of the universe that happened about 13.7 billion years ago. Nevertheless, referring to the authority of Einstein, some physicists want to persuade us not just that the apparent passage of time is an illusion, but even that the notion of time lacks all scientific foundation! So an urgent question of today is this: How can we trust a science which not only denies what everyone can observe with his own senses from instant to instant, but which also is in blatant conflict with itself?

2.   EINSTEIN ON TIME

In order to throw new light upon these problems we have to reconsider the ideas about time in physics which dominate the heritage from Einstein. How was Einstein's own view? To the philosophical question: *what is time?* he responded as a scientist by ignoring philosophy, appealing to what we see by observing a watch, viz., *clock-readings,* numbers marking instants. This answer, of course, is just as ingenious as it is natural and simple.

However, his answer says nothing about the difficulty of deciding whether the observed clock shows the right time. Still less does it inform us of the crucial problem which is this one: how do we distinguish a clock that works from one that does not? Of course this is a condition for deciding if a clock tells the right time, since a standing clock is right once or twice a day. But in consequence of his response, as mentioned above, Einstein simply determined the time of an event as the number immediately read off a clock associated with an observer.

From his reflections as described in several publications one can, however, deduce the following distinction: 1) if an observer is close to an event, his reading represents its *local time,* 2) if an observer is far from an event, his reading represents its *time at a distance,* which can be compared to its local time by a calculation taking account of the finiteness of the speed of light. That the speed of light seems to be independent of the motion of its source is here decisive.

Scientific objectivity hinges on the independent verification of data by several observers, so Einstein assumed that observers agree to calculate time at a distance using the same method. He furthermore prudently stressed the natural fact that all rational comparison of clock-readings, or epochs, necessarily presupposes that the clocks concerned go at the same rate [1920 ch.viii]: "It (is) assumed that all these clocks go at the same rate if they are of identical construction".

The statement quoted is notable for other reasons than its frank admission of time's flow: it was presented before the definition of simultaneity, but after the introduction of the famous train experiment where train and embankment are struck by two lightnings; and the succeeding argumentation aims precisely at showing that if the two flashes are taken to be simultaneous with respect to synchronous clocks at rest on the embankment, they cannot also be supposed to be simultaneous with respect to synchronous clocks co-moving with the train, and *vice versa.* The clocks, being of identical construction, are assumed to be pairwise synchronized.

*Therefore the special theory of relativity, including its suspension of the classical notion of simultaneity, explicitly relies on these premises concerning clocks in motion that not only do they go, thus indicating **the flow of time**, but being identical they even keep **the same rate**.*

From such premises, however, Einstein drew the conclusion that correctly synchronized clocks will mutually appear retarded relative to each other, and that correctly calibrated rods will mutually seem contracted relative to each other, all in consequence of their relative motion. From Einstein's interpretation of the Lorentz transformations *(LT)*, making up the mathematical contents of his special relativity, it follows that the contraction will vanish when the motion is brought to a halt, whereas the retardation, being one-sided, leads to an absolute effect.

We are therefore confronted with an obvious paradox: a theory which is explicitly based on the premise that all clocks involved are of identical construction, thus keeping the same rate, entails that clocks in relative motion do not count the same intervals of time after all!

Mogens True Wegener

Mathematically the theory is *consistent*. Physically it is *supported* not merely by a lot of experiments, but even by the most diverse kinds of experiment. It therefore seems impeccable. Nevertheless it is *mandatory to make the following reservations*: 1) No theory contains more truth than the premises it is based upon; so a mere lack of contradiction is a necessary, but not a sufficient, condition for its truth. 2) Although a theory can be falsified conclusively, it can never be verified conclusively; therefore no kind of observational support should ever be assumed to exclude the possibility that other premises might lead to even more plausible consequences.

How do we cope with the fact that a seemingly consistent theory is so full of paradox? In order to grasp this we must first probe for a deeper understanding of Einstein's theories. Making this effort we shall find a serious ambiguity inherent in his definition of time.

SR (special relativity) is based on two pillars: (1) **the relativity principle**, stating the impossibility of determining absolute motion or rest; (2) **the light speed principle**, stating the universal constancy of light speed, independently of the motion of its source. According to the relativity principle, all inertial observers and their reference frames are equivalent with respect to a scientific description of the laws of nature. According to the light speed principle, light is transmitted with equal speed in all inertial reference frames, irrespective of its direction. The theory is termed 'special' because it is restricted to the consideration of observers in inertial motion assumed to be uninfluenced by acceleration or gravitation. **GR** (general relativity) is also based on (3) **the equivalence principle**, stating the identity of acceleration and gravitation.

In both theories reference frames are in focus. What do we mean by a frame of reference? Of course, *reference frames* are *abstract entities* and *therefore immaterial*; they are conceptual constructs not inhering in nature. However, as their geometrical structure is influenced by the presence or non-presence of matter, they are often treated *as if* they had a material existence. The general theory is reducible to the special theory if the influence of gravitation is negligible; this shows the construction of inertial frames to be prior to the construction of accelerated ones. But the construction of inertial frames involves the notion of time at a distance.

In his famous paper to *Annalen der Physik* [1905 I §1], Einstein made it clear that our concept of time at a distance, or simultaneity, is based on definition. Now a definition is chosen because it is convenient and serves a scientific purpose; hence definitions have no truth value. But the definition of simultaneity is important, indeed fundamental, since, according to Einstein, "all our judgments in which time plays a part are judgments of simultaneity".

How did Einstein define his concept of simultaneity? He imagined two observers, each of them provided with his own clock, being involved in a steady exchange of light signals, each signal containing information about the clock-readings read off the other observer's clock. Granted that a signal goes zig-zag between the two observers, its reflection being instantaneous, the light speed principle seems to imply that its journey out and home must be of equal duration. Therefore it appears rational to define the epoch of reflection as half the sum of the epochs for emission and reception of the reflected signal, that is, as the arithmetic mean of these epochs. This, precisely, is the famous Einsteinian *definition of simultaneity*, or time at a distance.

In his 1905-paper, Einstein wrote: "We assume that this definition of synchronism is free from contradictions and possible for any number of points", adding these further assumptions: 1) If clock *A* is synchronous with clock *B* then, likewise, clock *B* is synchronous with clock *A*. 2) If clock *A* is synchronous with clock *B*, and clock *B* is synchronous with clock *C*, then clock *A* is also synchronous with clock *C*. Thus *clock synchronism* is both *reciprocal* and *transitive*.

In a later discussion (1952, ch.viii, footnote) he commented on the transitivity assumption: "This assumption is a physical hypothesis about the law of light propagation; it must certainly be fulfilled if we are to maintain the law of the constancy of the velocity of light in vacuo".

Einstein thus openly admitted that *the light speed principle is intimately connected to the definition of clock synchronism; in fact so intimately that the principle makes no sense except if interpreted on the basis of the transitivity of simultaneity.* Now it is precisely the transitivity of simultaneity which is jeopardized in consequence of **SR**. The point is that, following Einstein, the Lorentz transformations should be interpreted so that the *transitivity* of simultaneity holds good only *within* each single inertial frame, whereas it does *not* hold for comparison *between* inertial frames. But from the footnote just quoted it would seem to follow that the light speed invariance is only valid for light propagation *within* a single inertial frame, whereas it is invalid for the comparison *between* inertial frames. This further seems to entail that the light principle can be applied only to a single reference frame at a time, not to several frames simultaneously. Thus data referring to more than one frame at a time seem to be incomparable!

Einstein had no doubt that the light speed must be the same with respect to all frames. Apparently he never confronted the issue whether *the same photon* (whatever that is?) retains the *same speed* relative to *two different frames* at the same time (quite a feat!), or whether its speed is only constant relative to that frame wherein it is *observed*. This can hardly be decided by appeal to principle. But the situation is going to be still worse, before it can get any better. Of course, Einstein did not relinquish synchronizing the master clocks of different observers. No, he tacitly synchronized such master clocks in another way, by a different method!

Physicists need to compare data that refer to different inertial frames in relative motion. According to Einstein, the possibility of comparing clock readings from various inertial frames is guaranteed by the identical construction of the clocks involved. From the light speed principle it then follows that their measuring rods can be properly calibrated so that we can speak of their meters being identical. But this is not sufficient for a comparison of such frames. The observers must further agree on *a certain event* as signifying the *common time zero* for their calendars. If the observers ever meet, the event of coincidence will be their natural choice. If they do not, one of them could elect an *alter ego* at a fixed distance who does meet the other.

We will now disclose the ambiguity in Einstein's definition of time. It turns out to be connected with the conventionality of *the choice of time zero* and is reflected in the fact that, whereas we can speak of the instantaneous *coincidence of observers*, using this as a common time zero, we can only speak of the *coincidence of frames* if it is not momentary, but permanent. This is why we must distinguish between *proper time*, as read off from *the master clock* of a particular observer, and *frame time*, as shown by *some slave clock* fixed to his comoving frame: *master clocks of observers* are synchronized differently to *slave clocks within frames*!

The ambiguity inherent in Einstein's definition of distant time, or simultaneity, is this: While he **explicitly** defined **simultaneity within frames** by taking **the arithmetic mean** of the epochs of emission and reception of a reflected light signal, using the *master clock* of a single observer as a standard for calibrating the readings of his comoving cloud of spatially distributed *slave clocks*, he **implicitly** defined **simultaneity between frames** by taking **the geometric mean** of the same epochs, in order to calibrate the readings of the *master clocks* of different observers; in fact, "a frame" is a "cloud" of "clocks" associated with an observer, cf. Popper [1963].

Mogens True Wegener

Small wonder, then, that the slave clocks at rest in the comoving frame of one observer are advanced relative to the master clock of another observer! Why not state the facts this way? The master clock of a particular observer *may well appear* to be retarded with respect to the slave clocks that are fixed to the comoving frame of another observer. But that, of course, does not entail that the master clock of the first observer will deviate from that of the second one! Comparing the **master clocks** of two **fundamental observers** we find **no retardation**: if properly synchronized, the master clock of a fundamental observer can never be retarded relative to the master clock of another fundamental observer if their relative motion be inertial and collinear. That this is so is **a simple consequence of the definition of fundamental observers**: if the clock of one observer deviates from that of another, at least one of them is not fundamental!

A *general theory of time keeping* was outlined by E.A. Milne [1948], in collaboration with G.J. Whitrow; cf. the analyses in Stephenson & Kilmister [1958], H. Törnebohm [1963], and S.J. Prokhovnik [1967]. Their method, the so-called radar method, is summarized in the so-called *k*-calculus of H. Bondi. The *k*-calculus can be used to show that the observed retardation of moving inertial clocks originates from different definitions of simultaneity, different methods of synchronization. Such methods, just like the definitions upon which they are based, are purely conventional, as stressed by H. Poincaré, precursor of Einstein, and the first to invent the **SR**-formalism; cf. E.T. Whittaker [1953], G.H. Keswani [1964], and H.A. Lorentz [1921]. We conclude: **the claim that moving clocks are slow is based on a convention**, nothing else! The problem is altogether different, however, if the clocks involved are no longer inertial.

The fact that mesons from cosmic radiation can penetrate far deeper into the atmosphere of the Earth than possible, if the product of their mean life times with the speed of light is taken as a limit, in itself shows nothing about meson life-times being prolonged due to a relativistic retardation of their "internal clocks". When considered in isolation, such observations can just as well be explained by a Newtonian theory allowing the occurrence of super-luminal velocities. To this it can be objected that the observations must naturally be interpreted in the light of our knowledge that the speed of light constitutes a natural limit to the velocities of material bodies.

But to this it can be replied that the speed of light can be a natural limit to all signals, including moving mesons, without its numerical value being kept invariant between frames. The point is that we need to make a further distinction between **the one-way light speed** relating to its propagation between a source and a sink, and **the two-way light speed** which is the mean light speed, to and fro, for a reflected radar signal. It is indeed easy to see that a variable one-way light speed does follow directly from **SR** if the theory is rewritten in terms of Törnebohm's absolute coordinates, although he did not do the job himself; and by the same token all traces of time dilatation and length contraction can be eliminated (cf. this book ch.1, app.).

For my own part, I have made a hard attempt to work out a new theory of relativity in analogy with Törnebohm's and Prokhovnik's non-standard interpretations of **SR**. My idea was to combine a variable one-way light speed with a constant two-way light speed and, in line with this, to invent *a new unitary expression for time at a distance which in some cases is reducible to the arithmetic mean and which in other cases is reducible to the geometric mean*; cf.ch.s 3-4. This theory pulls the teeth out of de Sitter's proof for the one-way light speed independence of the proper motions of binaries. But, like **SR**, the theory may be unable to explain the experiment of Sagnac. Thus one should probably rather consider the possibility of returning to the theories of Poincaré and Lorentz; cf. J.S. Bell [1986], and F. Selleri [homepage R38].

Non-Standard Relativity

3.   THE BRITISH TRADITION

The British Tradition in relativistic cosmology is represented by the names of E.A. Milne, A.G. Walker, and G.J. Whitrow. They were all mathematicians, as shown by their approaches. Milne created the theory of *kinematic relativity* which was developed expressly as an alternative to the relativity theories of Einstein. His ambition was to construct a mathematical cosmology by formal deduction from a few definitions and principles; and the outcome, which is extremely ingenious, exploits the radar technique by generalizing *SR* into a full fledged cosmology.

Walker developed kinematic relativity further, so that the theory was no longer solely associated with Milne's world model, which is one of uniform dispersion from a transcendent point-event, a kind of imaginary "big bang". Like *GR*, it then became a mathematical technique applicable to a whole range of world models, viz., all those that are subject to cosmic isotropy. The *Robertson-Walker metric (RWM)*, incorporating the *principle of cosmic isotropy* formally, is still the most significant instrument of modern mathematical cosmology.

Whitrow gave important technical contributions to the development of the radar method. During a longer period he worked on kinematic relativity as an assistant and collegue of Milne. He later became renowned for his inquiries into the history and philosophy of time which are reported in his monumental *The Natural Philosophy of Time,* a work that turned out to provide inspiration for the founding of *The International Society for the Study of Time* (*ISST*, 1971). All members of the "kinematic league" were inspired, directly, or indirectly, by Poincaré.

In 1905, some weeks before Einstein published his *SR* in the *Annalen der Physik*, the great French mathematician and philosopher of science Henri Poincaré published an equivalent but formally more advanced theory in the *Comptes rendus*. In this paper he pointed out that the invariance of the speed of light enables us to unify time and space by defining spatial intervals in terms of temporal ones, and so he foreshadowed that the idea of *light-time* would replace the use of a "rigid rod" as the standard of distance. This prophecy was fulfilled with the invention of the *radar technique* in *GB* shortly before *WW2*. In another paper he suggested that the temporal coordinate does not represent "true time"; so time may, after all, be universal.

The significance of *the radar principle* is unique; not only is it used in nature by bats and dolphins: today we measure the radar distance to a planet with a precision down to centimeters. The *meter* is now defined as the path travelled by photons *in vacuo* in 1/299792458 sec; further, the *second* is defined as 9192631770 periods of the radiation corresponding to the transition between two hyperfine levels of the groundstate of the caesium 133 atom (*CGPM* 1983/1967). The theoretical importance of these definitions far overshadows their practical one: by stripping the presumed fundamentality off the idea of space, they flout the view of Einstein [1920 app.v] that "(physicists) endeavour in principle to make do with 'space-like' concepts alone".

In 1929, the astronomer E. Hubble discovered a systematic redshift of spectral lines in the light from distant galaxies. This redshift is traditionally interpreted as a sign that the galaxies are all receding from each other with velocities approximately proportional to their distances. Unless our galaxy is so privileged as to occupy the very center of the universe, this discovery entails that the system of galaxies is not static, but rather subject to a universal dispersion that has no definite center, since any galaxy, or galaxy cluster, can be said to constitute its center. This dispersion is most often described as an "expansion of the universe", but should rather be described as a simultaneous and proportional expansion of all distances between galaxies.

In 1965, the physicists Penzias & Wilson observed a strange radio noise produced by a smooth *cosmic background radiation*, **CBR**, coming from the most distant parts of the universe. The radiation, having a temperature of ca. 2.7 Kelvin, showed a so-called black-body spectrum. This fact has been interpreted as proof that the universe originated about 13.7 billion years ago in a huge explosion, the "big bang". But there are other ways of explaining this observation, such as re-radiation of stellar light from interstellar whiskers of carbon (graphite), or radiation emitted spontaneously from a so-called zero point field; cf. Narlikar [1980]. What may be said for sure is that both observations clearly show universal space to have no privileged directions. This fact puts great importance to **RWM** as the standard metric of the universe.

Milne died in 1950 without accepting **RWM**, but this does not make his work obsolete. As a nominalist he realized that reference frames are not real things or entities in nature, but abstract constructions of the human intellect; so he asked the question how to construct them. He felt foreign to the pseudo-metaphysical idea of Einstein that there is a natural or immanent structure of space. Instead he followed the lead of Poincaré who saw geometry as a construction of the human mind made to the purpose of the co-ordination of data, the numerical results of observation or experiment. It is here natural to make a comparison with the choice of geodesic projection which aims at obtaining the most convenient description of the landscape on a curved surface such as that of the Earth. Our choice of geometry is free, but it turns out to restrict our possibilities of description in definite ways. *Reality is not to be found in the spatial frame itself, but in the observational data which we insert into the frame.*

The German philosopher and multi-genius Leibniz took physical reality to be analysable into some metaphysical entities called monads. In his book *Cosmologie du XXme Siecle* [1960], J. Merleau-Ponty compared Milne's world-model to *a monadology translated into mathematics*. According to Milne, an observer like a monad can be viewed as a temporal series of experiences or events; but he abstained from spatializing this series into a world-line in so-called spacetime. Any single observer he assumed to be provided with two instruments: a clock, and a theodolite.

Milne's definition of a *clock* was mathematical: a one-one relationship between the series of events constituting the observer and an ever increasing series of numbers. That an observer is provided with a clock, therefore, to him just meant that he is *able to count* his own experiences. This puts Milne in company with A. Mercier according to whom a *clock* is an instrument for the *counting* of time, *not* for the *measuring* of time. Since durations can be counted and compared, but not measured, duration cannot be fundamental. So, if questioned, Milne and Mercier would probably have rejected the metaphysics of H. Bergson which is based on the idea of *durée*.

It is clear that the idea of a mathematical clock sketched above does not bring us very far. So Milne attacked two basic problems of time-keeping: 1) How decide whether two clocks go at the same rate? 2) How determine whether, considering a cloud of clocks, all keep the same rate?

His method for their solution was based on the radar principle, discussed previously. Clocks keeping the same rate, their time-keeping being adjusted to a common time zero, were termed *congruent*, and the mathematical functions describing the readings of clocks connected by means of zig-zag signals sent to and fro between them were termed *signal functions*.

Milne's solution of his two problems can now be boiled down to these two main points: 1) Two clocks are congruent if their signal functions are reciprocal, or symmetric. 2) Infinitely many clocks are congruent if their signal functions are commutative; the latter property implies that, even though their mutual distances change, their relative proportions are preserved.

Non-Standard Relativity

One can imagine the universe to consist of a large, maybe infinite, number of equivalence classes, each one made up of a large, maybe infinite, number of mutually equivalent observers. One could even imagine each pair of equivalence classes to have at least one common member. However, it follows from Milne's two conditions of clock congruence that, if two equivalence classes have more than a single member in common, the classes in question must be identical. Hence we have to distinguish between two kinds of observers: equivalent ones whose clocks are mutually congruent, and non-equivalent ones whose clocks are not mutually congruent.

In fact, Milne held that the general structure of the universe is fixed by the existence of a single, unique, and privileged, equivalence class of observers, the *substratum*, whose members are termed *fundamental*, in contrast to observers who do not belong to this class and who may therefore be termed *accidental*. It is easily realized that this privileged substratum constitutes a universal reference frame for the description of rest and motion of particles.

To sum up: Milne claimed the structure of the universe to be dominated by a substratum, i.e., a privileged class of equivalent observers, or particles, whose clocks all keep the same rate; further, all relative proportions of distance between such equivalent observers are preserved; moreover, all directions in the substratum are equivalent, so that there is no privileged direction. This **idea of cosmic isotropy** constitutes the essence of the so-called **cosmological principle**. Albeit the principle is often ascribed to Einstein, even by Milne, it was first coined by Milne.

Milne took the principle to be valid for a specific world-model in which the dispersion of fundamental particles always takes place with uniform velocity. Walker generalized Milne's ideas by independently developing his own version of the **RWM** so that it remains valid for all world models subject to cosmic isotropy irrespective of their general scale functions of distance. Thus, if the **RWM** holds for a specific world-model, this model is subject to cosmic isotropy. The definability of a cosmic time turns out to be closely related to the principle of isotropy:

It follows from Milne's cosmological principle, when interpreted by Whitrow's argument concerning signal functions, that it is possible to define a universal time within any world-model fulfilling the principle. The same conclusion follows from Einstein's **GR** when interpreted in the light of **RWM** whose basic parameter is the very same cosmic time. The fact that a cosmic time is definable for all world models subject to cosmic isotropy testifies to their rationality. Hence, the principle of cosmic isotropy may be used as a means of excluding certain world models from consideration, such as the "rotating universe" of Gödel, or the "multiverse" of Smolin.

So, when an adherent of standard relativity says that the time read off the master clock of a certain fundamental observer is delayed relative to the time shown by the slave clocks distributed over the co-moving frame of another fundamental observer, he is perfectly right; nevertheless he is missing the crucial point which is that, if the master clocks of fundamental observers are synchonized correctly, they will show the same time, viz., cosmic time; and, if they do not show the same time, the conclusion is that they are not properly synchronized.

As Einstein admitted: identical clocks, if exposed to identical forces, keep the same rate!

=//=

Mogens True Wegener

4.   ONE UNIVERSE ONLY

Taking nature to be governed by laws, it is the task of natural science to map these laws. Only theoretical connections sustained by observation and experiment can pass as natural laws. Every observer must construct an exact reference frame for the co-ordination of data, results of observation and experiment, before these can be communicated to other observers. Objective science presupposes the ceaseless exchange of information between equivalent observers who agree on common rules ensuring that their communication be not marred by inconsistencies.

The rules of communication in physics are rules for the transformation of co-ordinates. It is remarkable that such transformation rules may determine the very form of nature's laws. In fact, the basic law $\Delta E = \Delta mc^2$, for instance, is derivable from the Lorentz transformations. The demand for consistency exposes the possible form of natural laws to severe restrictions. However, *the very gist of the preceding paragraphs is that a proper interpretation of the Lorentz transformations presupposes that they be inserted into a cosmological context.*

In this way the idea of **cosmology** as *the science of the universe* is brought to the fore. Cosmology can be defined as the science of the universe as a whole, i.e., as the general science which brings all scientific disciplines into play in order to encompass the concept of everything. The difficulty, however, is that the universe shows itself neither to our reason nor to our senses. The universe in itself is a mysterious X which remains *incognito*, forever unknown to us.

The great German philosopher Kant, as we know, distinguished *reality* from *appearance* or, as he preferred to say, *things as they are in themselves* from *things as they appear to us*. This distinction is clearly relevant to our understanding of the universe as a totality of all things. We believe to be in immediate contact with unfeigned reality due to observation and experience. We think that we are able to form a concept about reality-in-itself, and in a sweeping manner we even speak about the world, or everything, identifying the world with the essence of facticity.

But it is not that easy, for in his *Critique of Pure Reason*, 1ˢᵗ antinomy, Kant showed that human reason entangles itself in contradictions when attempting to gauge everything at once, inevitably presenting the universe as both finite and infinite in both time and space. Kant was so proud of having shown his antinomies to be the limits of all possible experience that, in his *Prolegomena*, he declared himself willing to stake his entire philosophy at this single point. Therefore *it is interesting to notice that a very simple world-model, viz., Milne's, constitutes a counterinstance to this claim of Kant by showing his 1ˢᵗ antinomy to be dissolvable.*

The point is that Milne's model can be represented mathematically in two different ways: 1) as based on the t-scale, the frame time of fundamental observers using atomic clocks as their master clocks, relative to which such observers *recede* from each other with constant velocities, implying the subtratum to be in uniform *dispersion*; 2) as based on the τ-scale, each τ-value being found from the logarithm of the corresponding t-value, relative to which all fundamental observers are *at rest* but their atoms steadily *shrinking*, implying the substratum to be static.

So *a finite past* starting at $t = 0$ does not exclude *an infinite past* going back to $\tau \simeq -\infty$! From the mathematics it furthermore follows that the substratum which, according to t-time, can be mapped in *flat 3-space* as *a sphere with finite radius $r = ct$* in spite of its infinite contents, according to τ-time by contrast makes up *the static contents of infinite hyperbolic 3-space.*

Non-Standard Relativity

So, mathematically, Kant's so-called antinomy is free of contradiction: everything fits together! But, of course, the difference between reality and appearance is not thereby suspended.

Kant claimed that we can never know *reality-in itself*, only *reality-for-us* can be grasped. However, by successive constructing and eliminating *universes* in the sense of *world-models*, we can obtain approximate knowledge of the laws of nature and the structure of the universe. From this point of view cosmology assumes a central position among the natural sciences.

It was said that Milne's cosmology can be viewed as a kind of Leibnizian metaphysics translated into mathematics. Leibniz is famous for having developed the idea of possible worlds, an idea which today has become a topic of major importance in formal logics and semantics. One might wish that the proponents of the "many worlds" interpretation of quantum mechanics would have spent a little more time to make themselves acquainted with modal logic; this might have saved them from indulging and persevering in the worst of their extravaganzas.

Now it is natural to connect the idea of a possible world with the idea of a world-model. In this way a close bond is disclosed between classical metaphysics and modern cosmology. The function of a world-model is to map the structure of a certain class of possible worlds. While the model in an abstract way maps the formal properties of some given class of worlds, any single member of that class embodies a temporal succession (history) of observable events. So a possible world can be imagined as a closed succession of events, or a finished process.

A scientific world-model, accordingly, is the formal result of an attempt to map the laws valid for a certain class of possible worlds, or an attempt to decode the structure of these worlds. Hence identity of form and contents, structure and existence, can only be realized if the form constitutes its own content, i.e., if the structure poses its own existence; but this is impossible, Leibniz insisted, the only exception being the essence of God which entails "his" existence. With this argument he would have refuted Hawking as clearly as he refuted Spinoza.

Plato said: there is **one world only**, and "it is and remains the only one" (*Timaios 30^d*). Leibniz agreed: we can imagine an infinity of possible worlds, but only a single one is real. Now, what is *possibly true* can be understood as that which is true in some possible world. Likewise, what is *necessarily true* can be interpreted as that which is true in all possible worlds. Further, what is true in fact, or *actually true*, is what is true in the one and only actual world. We may add that what is *true in terms of natural law* is what holds good within a certain class of possible worlds on the assumption that they share the structure of a certain world-model.

What is wrong with the many worlds interpretation of **QM** is that its adherents confound this by insisting that all worlds produced by the process of quantum bifurcation are equally real. The result of such nonsense is what logicians have termed: the collapse of modal distinctions. But if it is said that the Ψ-function describes a probabilistic necessity that is equally applicable to all possible worlds for which **QM** holds, then nothing is wrong, yet nothing explained.

A possible world wherein nothing at all ever happens involves a contradiction in terms. What happens we call events, and events always take place in time or rather: they make up time. Consequently *each possible world*, including the one and only real one, should be understood as a temporal world course, i.e., as *a linear succession of events* forming an unbroken process. *That the actual world is one thus means that there is in fact a single all-comprehensive time.* And the fact that events happen, which is the nature of factuality and the factuality of nature, indicates **time's passage**, whence our partitioning of time in **past**, **present**, and **future**.

The *principle of the unity of the world*, postulating that there is in fact one world only, excludes "inflationary bubbles", thus eliminating one of the most cherished recent fantasms. Why speculate on that which must forever remain outside the limits of all possible experience? The principle that there is one world only entails the *postulate of a single unique world time*. By keeping all investigation within the limits of possible experience it recommends itself, the onus of proof resting upon the shoulders of those who would call it irrelevant or inadequate.

Granted that the notion of a possible world includes the notion of a unique world time in the sense of a linear course of events, the question arises how to understand such time properly. In order to answer this we have to exploit the methods of analysis offered by temporal logic. C.F. v.Weizsäcker [1985]: "A systematic reconstruction of physics would necessitate that a full calculus of temporal propositions be developed and utilized as (its) basic foundation".

Indeed, *time's passage can be analysed in terms of tense logic*, cf. Wegener & Øhrstrøm [1996] and the kind appreciation by J.R. Lucas [1999]. According to tense logic, time exists in the same way as clock-readings, viz., as numbers indicating the successive occurrence of events; that in itself does not imply "the real existence" of time or of instants. The old quandary posed by the question: "how fast, then, does time run?", allows of a simple answer, however, since the rate of, e.g., an atomic clock can always be estimated relative another atomic clock.

It is a simple fact which stands beyond any reasonable discussion that the task of science is triple, namely: 1) to *describe* the present, 2) to *explain* the past, and 3) to *predict* the future. In its very *raison d'être*, science presupposes this partitioning that makes it meaningful to speak of the passage of time, hence also its direction; and if some of its practitioners afterwards try to convince us that time is an illusion, and that talk of time's passage is nonsense, they should be dismissed with the message that they have totally misunderstood their own business.

It is flatly unacceptable that the scientific establishment should feign the production of so-called "knowledge" that everyone can see is incompatible with its own premises!

The principle of the unity of the universe (one would think that this was already latent in the very notion of an universe) also yields a solid basis for drawing some other far-reaching consequences. The bond of unity, consisting in the participation of all fundamental particles in the common time of the substratum, prevents the universe from being split up into enclaves delimited by horizons. The postulate of a *cosmic time* is therefore equivalent to the postulate of the absence of horizons, anticipated by Milne [1948] with his *no-horizon postulate*.

As already mentioned, a cosmic time is conditioned by the preservation of all proportions of distance in a substratum of equivalent observers, implying all directions in this substratum to be approximately equivalent, so that the universe, as predicted about 1450 by Nicolaus Cusanus, the principal source of almost all the ingenious ideas of Giordano Bruno, can be compared to: *a sphere having its center everywhere and its periphery nowhere.*

To this we may add that the cosmic sphere does not allow of any division of its contents. Not only will the universe emerge in approximately the same way to all fundamental observers, but the observable part of the universe, in principle, coincides with the entire existing universe: nothing is outside, it is all visible if our sight be sharp enough (cf. this book ch.1, p.21, lin.10f.). Consequently, the universe must constitute a closed totality, similar to a so-called "black hole". As Plato said: "Nothing comes into it and nothing goes out from it, for it has no outside".

Milne did not view his t-time, but only his τ-time, as a genuine world time. In line with this he was reluctant about Walker's attempt to introduce an all-comprehensive universal time. But this need not bother us if we remember that Milne's t-scale, representing the frame time t displayed by the comoving slave clocks of single observers, is wholly different from Walker's cosmic T-scale which is identifiable with the proper time shown by their master clocks.

As already hinted at, the mapping of the substratum in Milne's model differs according to whether it is described relative to t-time (*private frame time*) or to τ-time (*public proper time*). Walker's metric enables us to replace the many private t-scales with a single public T-time. But the τ-scale was already public. In fact, any world model subject to isotropy allows us to define an infinite number of public time scales. Of these, two are important: T-time & τ-time. So Milne and Walker both dealt with two privileged time scales: Milne t & τ, Walker T & τ.

According to Walker's T-scale, which he identified as the public proper time read off the master clocks of all fundamental observers, members of the substratum, the spatial extensions of the atoms composing all bodies in the universe remain invariant, while the relative distances of fundamental observers vary according to the same function of T. But according to his τ-scale, fundamental observers are at rest while their atoms shrink. As Eddington once aptly remarked: "The theory of the expanding universe is equivalent to the theory of the shrinking atom!".

The idea of a *cosmic time* can be seen as a corollary to the principle of *cosmic isotropy*. This conclusion has far-reaching consequences for the interpretation of the Lorentz formulae: if a unique all-comprehensive time is at all definable, then why not regard it as "the true time"? This time is the public proper time of all fundamental observers belonging to the substratum and can be read off all atomic clocks that are permanently at rest relative to the substratum.

A slave clock comoving with, thus keeping a fixed distance to, a fundamental observer, cannot coincide with that observer. Hence we shall have to distinguish *accidental observers* from *fundamental observers*, realizing that the clock readings of accidental observers always deviate from the *cosmic time* that constitutes "true time". This deviation hints at the possibility of explaining gravity by time. In fact: *gravitation is time*; cf. Mercier [1979].

In Milne's cosmology, the cosmic substratum functions as a *compass of inertia* (Weyl) marking all local deviations from the universal dispersion of matter. The arbitrary motion of an object relative to the substratum is therefore completely described by two pieces of information: 1) its velocity with respect to that fundamental particle with which it momentarily coincides; 2) its distance to that fundamental particle relative to which it is momentarily at rest, the first being its velocity in the substratum, and the second being its displacement in the substratum. Together they inform us of the object's deviation from *cosmic symmetry*.

The motion of an arbitrary material object in the substratum is also influenced by another asymmetry, viz., the deviation of other objects, mainly nearby ones, from universal symmetry. In order to describe the influence on an object by its local surroundings, Milne used an inverted Boltzmann equation describing the accelerations in a statistical ensemble as a function of their distribution; thus he succeeded in effecting *an ingenious reduction of gravitation to inertia*. Einstein, by contrast, spent a major part of his life by attempting to reduce inertia to gravitation; but to cram forces into the package of curved spacetime is a far cry from explaining them!

5.   SUGGESTIONS

As stated by Phipps, 1986: what is relevant of Einstein's **SR** is comprised in the γ-factor. However, this factor does not imply that differentials of proper time are "inexact", as opined by Phipps; there is no reason at all to suppose that standard frame time should be "exact"; and the Cern evidence has no bearing whatsoever on the relationship between fundamental observers.

*Realizing the definability of a Cosmic Time we have every reason to reject Einsteinian **SR** and to insist that the γ-factor $1/\sqrt{1-v^2/c^2}$, the importance of which I do not dispute, cannot be applied to describe the retardation of one fundamental master clock relative to another since such clocks keep the same rate and can always be adjusted to show the same cosmic time.*

Now stellar aberration, as well as the Sagnac experiment, seems to indicate that light in some respects behaves as if it were transmitted by a medium, an ether. That this is so has been argued by Selleri and others. But an ether needs not be stationary, since it may be dissipating. In fact, the ether hypothesis is best interpreted in terms of a substratum of fundamental particles; understood thus, we have the ideas of McCrea and Prokhovnik (cf. this book, ch.1, p.21).

That light behaves *as if* it were transmitted by a substratum means that light exchanged between two accidental particles, source and sink, emitter and receiver, is transmitted *as if* it were exchanged between two fundamental particles, viz., that with which the emitter coincides at the instant of emission, and that with which the receiver coincides at the instant of reception. Such transmission can only be described as a process already completed. Considering how light is going to be transmitted, we shall need to make a calculation of future probabilities.

It is at once obvious that such peculiar "transmission by substitute" must presuppose that a common time is definable for the two fundamental particles that serve as substitutes, one for the emitter, the other for the receiver. The instant of emission at one place in the substratum must be univocally comparable to the instant of reception at another place in the substratum. If the transmission is instantaneous these two instants may even be simultaneous, i.e., relative to cosmic time, the proper time read off the atomic master clocks of fundamental particles.

Further it is clear that an exact location of the events of emission and reception, spatial and temporal, within the substratum must presuppose that the electromagnetic waves involved in the transmission of light have suffered a sort of "quantum collapse", the alternative being absurd: electromagnetic energy being transmitted in the substratum without ever being received. So a substratum theory would naturally interpret light in terms of "photons", understood as the instantaneous exchanges of quanta of action between emitter and receiver, source and sink.

We earlier mentioned two important representations of the substratum: 1) one in terms of *cosmic time T*, according to which the *radii of atoms* (the constituents of fundamental particles) are *invariant* by definition, whereas the distances between fundamental particles are expanding; 2) the other in terms of *cosmic time τ*, according to which the *mutual distances* of fundamental particles are *invariant* by definition, while the sizes of their atomic constituents are shrinking.

Now I do not deny that the instantaneous "light speed", i.e., the speed of light defined as a quotient between the spatial and the temporal units of fundamental observers, remains invariant. But does this imply that the average "light speed", as integrated over the tempo-spatial interval separating source and sink, emitter and receiver, must likewise remain constant? Of course not! So we have to consider the possibility of "photon speeds" deviating from unity.

The possibility of "light being stretched" over cosmic distances as a consequence of the so-called "expansion of the universe" has been discussed by a number of cosmologists.

Let us look at the **SR** formula for clock retardation as expressed by means of the γ-factor:
$$\gamma = dt/dT = 1/\sqrt{1 - v^2/c_o^2} \equiv 1/\sqrt{1-v^2}$$
with t representing standard *frame time*, T representing so-called *proper time*, and $c_o \equiv 1$ being the constant quotient between the standard frame elements: dr of space and dt of time.

In his monumental book on *The Natural Philosophy of Time* (1961), Whitrow suggested an analogy between the γ-factor of **SR** and the Robertson-Walker metric of modern cosmology. The analogy emerges if we make the identification $v \equiv dr/dt \equiv R(t)\, d\sigma/dt$, whence:

(1) $\qquad dT^2 = dt^2 - dr^2 = dt^2 - R^2(t)\, d\sigma^2 = invar.$

(2) $\qquad dT = 0 \quad \Rightarrow \quad \sigma \equiv \int_{t_1}^{t_2} dt/R(t) = const.$

Here dT is an invariant element of *timespace* (better: *supertime*, as proposed by Mercier), and $d\sigma$ is an element of distance as measured in the substratum where σ is a fixed ("comoving") coordinate of some distant fundamental observer, and $R(t)$ is a dimensionless scale function for such elements, often called "the expansion factor" of the substratum.

However it is clear that, by this analogy, dr can no longer be the element of standard frame distance; the reason is that the spatial grid of *a standard frame* keeps its dimensions *fixed*, as measured by atomic radii, while the spatial grid of *the universal substratum* does not since, by definition, it is *expanding* relative to the radii of atoms. But the comparison is interesting.

Furthermore, the **RWM** can be reinterpreted so as to allow for the *shrinking* of atoms:

(3) $\qquad c^{-1}(\tau) \equiv dt/d\tau \equiv R(t)$, whence eventually

(4) $\qquad dT^2 = dt^2 - R^2(t)\, d\sigma^2 = c^{-2}(\tau)\{d\tau^2 - d\sigma^2\}$

Thus the "expansion" of the substratum, compared above to a "stretching" of the average *integrated light speed*, corresponds formally to a secular reduction of the sizes of atoms, which can equivalently be described as a secular reduction of the momentary *differential light speed*. This secular reduction is describable by the function $c(\tau)$. Being instantaneous and ubiquitous, it is imperceptible, i.e., it cannot be observed locally by any observer.

For any pair of fundamental particles, $\sigma = const.$, whence $dt = dT$ which, with general agreement about an universal zero of time, would further lead to the identification $t = T$, NB. For accidental particles, however, we have $d\sigma \neq 0$, whence $dt \neq dT$; this deviation from true cosmic time may indicate the emergence of spontaneous accelerations, i.e., of "forces"!

Hence the "cement of the universe" is not its ubiquitous network of causal connections in "spacetime", but rather that *absolute, unconditional, all-comprehensive simultaneity* which distinguishes a true **Cosmic Time**. Such time, indeed, the basic independent parameter of **RWM**, is simply *indispensable to a rational science of the universe, i.e., a scientific cosmology*.

$$=//=$$

CHAPTER 3

MILNE'S KINEMATIC RELATIVITY
IDEAS OF COSMOLOGY: A PHILOSOPHER'S SYNTHESIS

Revised version (2021) of a paper printed in:
Duffy & Wegener, eds.: 'Recent Advances in Relativity Theory', vol.1
Hadronic Press 2000 (ISBN 1-57485-047-4)

1.   COSMOLOGY, A SCIENCE?

All science is cosmology!   Sir Karl Popper [1958]

In our time, ***Cosmology*** is generally acknowledged to be *the science of the universe*. But what is *the universe*? Is it being, entity, or substance? Is it nature itself, ultimate reality? How do we overcome the desperate difficulties of speaking sensibly of everything at once? And in what sense can such an elusive subject be the object of anything like real science? Can we avoid the danger of assuming either too little or too much even before we begin? Finally, the universe is one, or unique: how can it then give rise to a legitimate science at all? Such questions cry for their rational answers but, of course, it is easier to ask than to reply. Hoping for better progress later, we shall start by making our language a bit more precise.

Neglecting the traditional mythological and metaphysical implications of cosmology, which are better left to the humanities for investigation, we prefer to concentrate on those senses of the term 'universe' which are more often brought up in relation to modern science; we shall therefore primarily use the term in the plural, to betoken scientific world-models. Nevertheless, it cannot be ignored that we all, for the purpose of preserving our lives in spite of an often hostile world, develop for ourselves what we might call: a practical metaphysics. We thus mostly accept that humans are mortal beings, that nature appears to be governed by general laws and, hence, that what happens may be explained as the effects of natural causes. So we feel reasonably convinced that if we venture to jump into a vulcano (some like it hot!) we shall hardly avoid to burn up, and that if we kick a stone we shall probably hurt our foot. Thus we are tempted to assume the existence of a unique ordered entity: ***The Universe***.

Now, do such reasons entitle us to claim that *there is* an Ultimate Material Reality, that the Universe *must* possess a definite Formal Structure, that the Course of Nature *must* be ruled by Law, and that the Principle relating Cause and Effect *is* valid without exception? These questions are not an issue of science, nor of common sense, but of philosophy; and to answer them in the affirmative would presuppose that our own practical metaphysics which is private, or particular, does in fact entail a theoretical metaphysics which is public, or universal. For people with an innate distaste for ontology this consequence may appear rather appalling: obeying some rules of behaviour clearly differs from adopting their casual conceptualization. Without committing myself to the entire critical philosophy of Kant, I must confess that his famous distinction between *reality-for-us* and *reality-in-itself* seems to be of relevance here, practical metaphysics relating to the first, and theoretical metaphysics treating of the latter. Armed with this distinction we are able to separate a *practical metaphysics* which is useful, even unavoidable, from a *theoretical ontology* which is redundant, and most often odious.

But are we not, for the purposes of science, in need of a very abstract and general notion of the universe, one which might serve as an ultimate instance of reference and whose function it is to promote the final unification of the disparate elements of human experience? It seems that we are, and here again we can benefit from Kant's transcendental philosophy. According to Kant, there are two types of concepts: *ideas* whose function is *regulative*, viz., to integrate our experience into a totality, and *categories* whose function is *constitutive*, viz., to differentiate the various kinds of our experience; with reference to the universe he speaks of cosmological ideas in the plural, but probably he refers to different aspects of the same. Now I do not want to let my own position depend on the right interpretation of Kant; however, I believe that I speak in accordance with the general tenor of his thinking when I propose the following *critical idea* of the universe, in contradistinction to the copious ideas of meagre value implied by the traditional profusion of pretentious metaphysical ontology.

The *critical idea of the universe* which I am here going to advocate is distinguishable from all ontological ideas of the same by the fact that it presents *the universe* as *an unknown X*, an unique, absolute and ultimate referent ("thing-in-itself") devoid of any specifiable properties. It serves the purpose of providing a minimum foundation for the stance of scientific realism, and in this way it seems *correlative to truth conceived as a regulative idea* in the sense of Popper. If a theory is to be true it must be true of an object which it represents and to which it refers, but as theories may be falsified, and never verified, we shall never be able to recognize truth; so it would be rash to hypostasize an ontology by ascribing definite properties to the universe. In the following I shall write *Universe* whenever I refer to some metaphysical idea, whereas I shall write *universe* in order to denote the idea of a totality devoid of intrinsic properties. With this convention I follow the proposal of Harrison [1981], the only difference being that he restricts the spelling with a capital U to the singular, saving the spelling without capitals for the plural, whereas I allow both spellings to be used both in the singular and in the plural.

My reason for this divergence is twofold: (1) The Universe of a metaphysical praxis needs no name as long as it is not deliberately conceptualized into some metaphysical theory. Now, although the proponent of a metaphysical theory will inevitably tend to believe that his Universe (capital U, singular) is the only true one, history knows innumerable examples of other Universes (capital U, plural) which are not merely different, but mutually incompatible. So the plural is needed too. (2) Although it is reasonable to identify different world-models with different kinds of universes (no capitals, plural), we shall nevertheless refer to a unique instance whenever a specific model is falsified by confrontation with experience and we want to make clear that this model does not properly represent the universe (no capitals, singular), and it is precisely to this purpose that the critical idea of an empty referent, an *X*, is needed. The advantage of the present approach is that we can retain a critical realism although we are incessantly reminded that even our best ideas and theories may turn out to be false.

I shall henceforth define **cosmology**, *the science of the universe* (in singular), as the art of inventing, or constructing, universes (in plural), i.e., as *the art of designing world-models*. It is an **art** because its practitioner, the constructor of world-models, must exercise *phantasy*. It is a **science** in so far as the free exercise of phantasy is kept within strict bonds stemming from *the obligation to obey the codes of scientific method* which are: (α) to pay due regard to formal *consistency*, and (ω) to respect the results of repeated *observation* and *experiment*. I shall further define a *world-model* as the material interpretation of a formal calculus which is taken to represent the structure of the universe. As we are forever unable to compare the *structure for-us*, viz., as *representation*, with the *structure in-itself*, viz., as *reality*, except in so far as the representation may be tested pointwise against experience, it is consistent to say that *the structure of the universe in a sense has no real existence except as representation*.

However, it must be admitted that *absolute identity is possible*, as identity of structure; this fact opens the hypothetical possibility that our cosmological theory could be a hit in the sense that the structure of the model materializing our theory might, eventually, be identical with the "real" structure of the "actual" universe. It should nevertheless be clearly recognized that, as soon as we depart from the *hypothetical* way of expressing ourselves in favour of a *categorical* manner of speaking, when we hazard ascribing a final structure to the universe, we transgress the limits of science and give way to free speculation. So we have this scheme:

Metaphysics: explains reality-in-itself
The Universe: an imaginary totality | *Other Universes: equally subjective*

Cosmology: describes reality-for-us
the universe: an unknown referent | *other universes: various world-models*

It is a commonplace to distinguish *theoretical* cosmology from *empirical* cosmology. The difference is mainly considered as being a matter of orientation; however, in the present context it is more illuminating to take a clear stand by presenting it as a question of principle. In this way our presentation offers a comment to Bondi (1961) who confronts inductive or extrapolating cosmology with deductive cosmology, i.e., cosmology founded on principles. For the sake of distinctness, we shall provisionally identify *empirical cosmology* as *inductive* and *theoretical cosmology* as *deductive*. By tradition, most cosmology has been inductive. The standard procedure has been to accept that the universe is governed strictly by so-called laws of nature and to build up models of the universe by generalizing those laws inductively, a procedure leading to questions about the relativistic invariance, or covariance, of laws.

Without denying the relevance of such issues it should suffice to recall the famous Bohr-Einstein controversy in order to confirm that the idea of deterministic laws dormant in nature is certainly not without problems; further, the idea of stochastic laws is not much worth after all, for how can we accept a "law" which is compatible with the most extraordinary deviations?

Moreover, most *empirical cosmology* is produced by a kind of "piecemeal engineering" that makes its structure resemble a patchwork of conflicting elements: thus, even when principles are introduced, it is hard to check their independence, and their consistence remains hazy.

I shall here venture to advocate **theoretical cosmology** as *the* physics of *the* cosmos, i.e., as a deductive science which, exploiting a formalism based on definitions and principles, considers *the ultimate structural totality of the universe* to be its natural object of research. This stance is not meant to imply that there is no place for empirical cosmology: giving the lead to theory as the motor of research makes the need for effective brakes the more urgent. Hence *the proper function of empirical evidence* is to act as a needful brake on theoretical speculation in order that it shall not forfeit its foundation in our ordinary human experience. For that very reason there is an intimate relation between our *practical metaphysics* and our *empirical cosmology*, both giving important contributions to our so-called "world-picture". Now, what can be said of *theoretical cosmology*, if we are to abstain from principles of reason? Not much, I am afraid, but that is no reason to be silent; cf. my [1994] (this book, ch.7).

So let it be said: *á priori it is a precept of natural science* that its aim be threefold, viz., **(1)** *to describe the present* in order **(2)** *to predict the future* and **(3)** *to explain the past*; therefore *the triple division of time into **past, present and future** is indispensable* to science. *It is simply a gross misunderstanding to believe that **TIME** - the basic feature of all existence - can be deduced from or explained by science, as it is a fundamental presupposition of science; but neither should we consent to those forms of science which are determined to ignore time. The least to be expected from **a science which is faithful to experience** is that its conceptual representation of nature remains compatible with the passage, hence also the direction, of time.* But precisely that is not the case with those models of the universe which represent reality as some multi-dimensional "continuum" existing timelessly, like a piece of crystalline mineral. That attempts to soften them up by introducing a *temporal order* have met with success should not obscure the fact that nothing less than the full acceptance of a *temporal flux* will do.

It is astonishing and one of the paradoxes of history that, faced with today's science, we shall have to insist on the factual evidence born out by everyone's own bodily senses.

2.   MILNE'S KINEMATIC RELATIVITY

The theory with its recent developments now covers not only cosmology but a great part of theoretical physics, and the extent of its achievements is greatly to be admired.
Sir Hermann Bondi [1961³]

One cosmologist, probably the only one of his days, who recognized the fundamental importance of time to natural science, was the British mathematician E.A. Milne of Oxford. His ambition was to present cosmology as a purely deductive system, without any loopholes.

As noticed by North [1965], at least some of Milne's more inflated claims were untenable, but he nevertheless clearly concedes that Milne "seems always to have hovered on the verge of a perfectly reasonable hypothetico-deductive account of scientific explanation" (p.303).

For example, Milne claimed to have promoted the law of gravitation from the status of an empirical result to that of a mathematical theorem. On account of the hypothetical elements which were latent in this deduction, his collaborator McCrea [1953] denounced this claim.

Mogens True Wegener

Nevertheless, as he also admitted: "Milne *did* show that, starting from only a few general hypotheses about the contents of the universe as a whole, it is possible to infer properties that have something of the character of physical laws" (p.337). However, this is not very surprising, according to McCrea: "We know from other theories that we can start with physical laws and construct model-universes; if desired, we can reverse the mathematics" (p.338).

Milne constructed an extremely simple model of the universe based on his **kinematic relativity** which was developed as a challenge to the special and general theories of Einstein. He took his point of departure in the fact that *simple perception* will suffice to convince all conscious beings of the *passage of time*: time is a necessary concomittant of all existence. The basic ingredient of his world-model, then, is an abstract image of the human conscious-ness which we shall here call a *particle-observer*, or an *observer-particle*, or simply: a *monad*.

That Milne's kinematic cosmology can be viewed as a Leibnizian monadology translated into mathematics was pointed out by the French historian of ideas J. Merleau-Ponty [1965]; so here is a case where metaphysics may have proffered inspiration to scientific cosmology. The monads of the kinematic universe are primarily characterized by a capacity for continual communication: they are supposed to exchange information mediated by signals all the time.

In what follows I shall offer my own interpretation of Milne's ideas, trying to avoid those pitfalls which have most often evoked a censure from scientists of a more traditional stance. My presentation will, to begin with, concentrate on the two basic problems of time-keeping: (*a*) that for a pair of particle-observers, and (*b*) that for an infinite set of particle-observers.

Although I cannot recall any immediate reference to conventionalism, Milne was surely a genuine conventionalist in the spirit of Poincaré. Thus, according to Milne, any mechanical device which associates events with a scale of ever increasing numbers may serve as a clock: a clock in this sense is a local one-one relation between events and the set of real numbers.

The *first problem of time-keeping* can be stated thus: Granted that a particular observer P has set up an arbitrary clock for himself, is it then possible for another observer Q to set up a clock congruent to the first one in some specified sense, irrespective of their relative motion? It should be clearly understood that the universe of discourse is here supposed to contain P & Q together with their clocks, and nothing else, so their motion is perfectly relative and symmetric.

The solution given by Milne assumes that signals exchanged between the observers are reflected immediately; they therefore constitute a single unbroken zig-zag signal. It further pre-supposes that all signals contain information of the clock-reading indicating the epoch of the latest reflection-event, so that observers are able to read off one another's clock indirectly.

Let us assume that a zig-zag signal corresponds to this series of clock-readings:

$$t_1^p \leq t_2^q < t_3^p < t_4^q < t_5^p < t_6^q$$

It is then possible for Q to plot his clock-readings against those of P, the graph giving rise to a so-called *signal-function* θ. In the same way it is possible for P to plot his clock-readings against those of Q, and this graph gives rise to another *signal-function* ϕ. We therefore have:

$$(1) \qquad t_{\nu+1}^q = \theta(t_\nu^p) \; . \Leftrightarrow . \; t_{\kappa+1}^p = \phi(t_\kappa^q)$$

Following Milne, the two master-clocks, $\boldsymbol{T}^p$ of P and $\boldsymbol{T}^q$ of Q, can then be made congruent, $\boldsymbol{T}^p \simeq \boldsymbol{T}^q$, if one of the observers, say Q, is able to *regraduate* (the readings of) his clock in a way that makes the signal-function ϕ identical with the signal-function θ, so that: $\phi \equiv \theta$.

The problem is reducible to that of finding the square-root of the functional product $\phi\,\theta$, and Milne, assisted by his collaborator Whitrow, demonstrated that this can always be done. The significance hereof is that two observers are *equivalent* in the sense that their clocks are *congruent* if and only if the signal-functions connecting the observers are fully symmetric.

So far, our universe of discourse contains only two monads, or particle-observers, and their equipment, i.e., the master-clocks and the apparatus required for signal-communication; there is no trace of any forces or fields, neither do we presuppose the signals to have a velocity. What we dispose of to construct our world-model is up to now only an abstract calculus of t-numbers interpreted as the clock-readings of two observers connected by a zig-zag signal, and our only assumption is that the involved reflection-events do not occur simultaneously. This is an extremely meagre universe, of course, and, in order that the comparison of our world model with experience shall make sense, we need to expand its structure somewhat more.

We therefore proceed to the *second problem of time-keeping* which can be stated thus: Given that a primary observer O has constructed an arbitrary clock for himself, are two other observers P & Q then able to set up congruent clocks irrespective of their relative motions?

Let the signal-function from O to P be θ_{op}, that from O to Q be θ_{oq}, that from P to Q be θ_{pq}, and let further the functional inverse of θ_{op} be θ_{op}^{-1}, that of θ_{oq} be θ_{oq}^{-1}, and that of θ_{pq} be θ_{pq}^{-1}. Assuming the observers to be equivalent, their clocks have to be congruent, which implies:

(2) $$\theta_{op} \equiv \theta_{po} \ . \ \theta_{oq} \equiv \theta_{qo} \ . \ \theta_{pq} \equiv \theta_{qp}$$

The simplest case to consider is that where the observers are collinear in a fixed order; this case Milne defined by the transitivity, in the given order, of the relevant signal-functions:

$$\theta_{pq}\theta_{op} \equiv \theta_{oq} \ . \Leftrightarrow . \ \theta_{qo} \equiv \theta_{po}\theta_{qp} \quad \text{corresponding to the order} \quad O,P,Q \Leftrightarrow Q,P,O$$
$$\theta_{qo}\theta_{pq} \equiv \theta_{po} \ . \Leftrightarrow . \ \theta_{op} \equiv \theta_{qp}\theta_{oq} \quad \text{corresponding to the order} \quad P,Q,O \Leftrightarrow O,Q,P$$
$$\theta_{op}\theta_{qo} \equiv \theta_{qp} \ . \Leftrightarrow . \ \theta_{pq} \equiv \theta_{oq}\theta_{po} \quad \text{corresponding to the order} \quad Q,O,P \Leftrightarrow P,O,Q$$

It follows at once that congruence of collinear clocks implies that signal-functions commute:

(3) $$\theta_{pq}\theta_{op} \equiv \theta_{op}\theta_{pq} \ . \ \theta_{qo}\theta_{pq} \equiv \theta_{pq}\theta_{qo} \ . \ \theta_{op}\theta_{qo} \equiv \theta_{qo}\theta_{op}$$

The general commutative identity $\theta_{pq}\theta_{rs} \equiv \theta_{rs}\theta_{pq}$ was found to have two solutions:

(4) $$\theta_{pq}(t) = \omega\{\omega^{-1}(t) + \lambda_{pq}\} = \psi\,e^{\alpha_{pq}}\psi^{-1}(t)$$

These solutions were now taken to define the concept of a *linear equivalence* of observers. It was also shown that, if two members of a linear equivalence ever coincide, they all coincide:

$$\theta_{qp}\theta_{pq}(t_1) = t_1 \ . \Leftrightarrow . \ \psi\,e^{2\alpha_{pq}}\psi^{-1}(t_1) = t_1 \ . \Leftrightarrow . \ e^{2\alpha_{pq}}\psi^{-1}(t_1) = \psi^{-1}(t_1)$$

If $P \neq Q$, then $e^{2\alpha_{pq}} \neq 1$, whence $\psi^{-1}(t_1) = 0$, and the conclusion follows by generalization.

The linear equivalence can be depicted as follows. Imagine a set of discrete particles separated by equidistant intervals along a string which is homogeneous and perfectly elastic. The distances of the particles will be proportional to the tension of the string, whence their relative motions will be dependent on how the string is slackened or tightened, and how fast. If the motion of a single particle is uniform, relative to a given observer-particle, that of all other particles will be uniform, and their distances will be proportional to their velocity.

Mogens True Wegener

Now the structure of the model must be extended to 3-space; the linear equivalence must thus be generalized into a **universal substratum** which is an equivalence in 3 dimensions.

Intuitively it is not hard to see that, if the substratum is not a continous fluid but constituted by discrete particles, they cannot be equally distant from their neighbours in all directions. The consequence hereof is that, in 3-space, small perturbations should be expected to occur which might imitate the newly observed minute ripples in the cosmic background radiation.

Using standard coordinates, Milne could compare the clock-rates of different observers. With distances defined by *radar-signals* the rigid rod becomes redundant as a measuring device. He therefore interpreted the rate of radar-distance traversed to time spended as a proportionality factor, which thus constitutes an alternative to the light-principle of Einstein.

In retrospect this suggests that the quantum properties of "photons" be taken seriously; the visualization of them as "something travelling" in space is probably misleading.

To Milne, the step to a construction of transformations of coordinates was now small. Assuming the linear equivalence to be in uniform expansion as measured by so-called t-time, a time-scale which he later identified with that defined by the intrinsic oscillations of atoms, he was able to demonstrate the correct transformations to be those of Lorentz and Poincaré.

The form of signal-functions being $\theta_{pq}(t) = e^{\alpha_{pq}}(t)$, they are found to be ($c = c_o = 1$):

$$t_2^q \equiv t' + r' = e^{\alpha}(t+r) \equiv e^{\alpha}t_1^p$$
$$t_4^q \equiv t' - r' = e^{-\alpha}(t-r) \equiv e^{-\alpha}t_5^p$$

(5) $$t' = t\cosh\alpha + r\sinh\alpha$$
(6) $$r' = r\cosh\alpha - t\sinh\alpha$$

(7) $$\theta = e^{\alpha} = \sqrt{\frac{t+r}{t-r}} \underset{r'=0}{=} \sqrt{\frac{1+v}{1-v}}$$

The interpretation of t-time as atomic time, when multiplied with the universal constant c (the light speed), may be used to define the sizes of atoms as invariant according to the t-scale. The double solution to the commutativity problem indicates the possibility of a second scale of time related logarithmically to the first one according to the formula: $\tau = t_o \log(t/t_o) + t_o$.

The τ-scale has the following peculiar property: If the clocks of all observer-particles constituting the universal substratum are instantaneously regraduated from t to τ according to this formula, they will find their relative distances to be constant with regard to τ-time. Further, if they have chosen their 3-space to be flat according to t-time, in agreement with the validity of the Lorentz transformations, they will find their 3-space to be hyperbolic with respect to τ-time.

Therefore the universal regraduation from t-time to τ-time of all clocks will transmute a uniformly expanding substratum in flat space into a stationary substratum in hyperbolic space. However, the structure of the substratum, and its internal proportions, must remain the same in spite of the regraduation. So all atoms must shrink continuously, according to the τ-scale.

We now introduce a crucial distinction in the model between two kinds of particles, viz., *fundamental particles* that belong to the substratum, and *accidental particles* that do not. Each fundamental particle is a universal center of spherical symmetry in the sense that its associated observer will see all the other particles of the substratum distributed isotropically. This crucial property is preserved irrespectively of whether the t-scale or the τ-scale is used.

What happens when an accidental particle is released in the presence of the substratum? We said that, if the substratum is mapped according to τ-time, it is perfectly stationary except for a steady decrease of the size of atoms (which we here ignore): no external forces are in play. So we conclude that the motion of an arbitrary accidental particle is perfectly inertial in τ-time.

Now it is intuitively obvious that the motion of a particle, which is inertial when mapped relative to the τ-scale, will no longer appear inertial when it is mapped relative to the t-scale. *The effect of passing from τ-time to t-time is that inertial motion is replaced by accelerated.*

What Milne demonstrated was in fact that, when the motion of a test-particle is analysed mathematically with respect to the t-scale, the result is the emergence of an apparent "force" giving the particle a spontaneous acceleration which depends on the degree of its asymmetry. *So his idea was to explain the nature of "forces" by local deviations from global symmetry.*

This brilliant program he proposed to implement by an exceedingly ingenious procedure. Having analysed the motion of a single test-particle in the substratum, he next examined the effects of asymmetry on a whole infinite set of test-particles superposed upon the substratum. To this purpose he inverted the famous Boltzmann equation of statistical mechanics in order to derive accelerations from a position-velocity distribution instead of the other way round.

His conclusion was that any test-particle, when exposed to the influence of the substratum as well as that of particles not belonging to the substratum, will receive an acceleration which is the resultant of two different contributions: (a) one directed towards the provisional center of the universe as perceived by the test-particle, viz., the fundamental particle with regard to which it is momentarily at rest, and (b) another directed towards the local center of the set of accidental particles superposed on the substratum. The formula for the induced acceleration becomes:

$$(8) \qquad d\boldsymbol{v}/dt = -(\boldsymbol{r} - \boldsymbol{v}t)(Y/X)\big(1 + C/\psi(\xi)(\xi - 1)^{3/2}\big)$$

C being a constant, ψ an indefinite function, and $\boldsymbol{r}$ & $\boldsymbol{v}$ vectors of position & velocity; further:

$$\xi \equiv Z^2/XY \ . \ X \equiv t^2 - \boldsymbol{r}^2/c^2 \ . \ Y \equiv 1 - \boldsymbol{v}^2/c^2 \ . \ Z \equiv t - \boldsymbol{r}\boldsymbol{v}/c^2$$

For small $\boldsymbol{r}$ & $\boldsymbol{v}$, with $\boldsymbol{l} \equiv \boldsymbol{r} - \boldsymbol{v}t$, $M \equiv M_o \frac{C}{\psi(1)} \equiv [\frac{4}{3}\pi(ct)^3\Delta]\frac{C}{\psi(1)}$ and $\gamma \equiv c^3t/M_o$, we derive:

$$(9) \qquad d\boldsymbol{v}/dt \simeq -\boldsymbol{l}/t^2 - C\frac{c^3t}{\psi(1)}(\boldsymbol{l}/|\boldsymbol{l}|^3) = -\boldsymbol{l}/t^2 - \gamma M\big(\boldsymbol{l}/|\boldsymbol{l}|^3\big)$$

By means of so-called "superpotentials", Milne also constructed a new electrodynamics. He likewise considered the nature of cosmic rays, the structure of galaxies, and that of atoms. He further proposed the use of the formulae discovered by W. Voigt for the transformation of coordinates between fundamental and accidental particles when the τ-scale is referred to.

In his last years he got problems as regards the correct relation between a t-time mapping and a τ-time mapping, especially when it comes to the representation of optical phenomena. He did not live long enough to solve these problems and - his ideas not finding the followers they deserved - with one exception, see §3 - his cosmology remained partially incomplete.

3.   WALKER'S ANALYSIS OF MILNE'S IDEAS

As Eddington once remarked: *The theory of the 'expanding universe' might also be called the theory of the 'shrinking atom'.*   Quoted from Whitrow [1961, p.248; 1980, p.293]

Mogens True Wegener

Despite the obvious importance of Walker's contributions to relativistic cosmology they suffered the bad luck of being overshadowed right from the beginning by the slightly earlier work of Robertson, and only a very few of the experts in the field appear to be acquainted with his original papers. It therefore seems pertinent here to outline some of their contents.

In a paper on Milne's kinematic theory [1936], Walker stressed that Milne's choice of the Lorentz formulae to connect the observational data in t-time involves an unnecessary restriction. He therefore set himself the task of investigating the possibility of applying Milne's kinematic method to models which do not presuppose the validity of the Lorentz formulae.

In order to perform this task, he assumed the following two principles to hold good: (1) the *cosmological principle (CP)*, in the formulation by Milne, implying that the totality of observations that a fundamental observer can make is "similar" (involving identity of structure) to that of any other fundamental observer; and (2) the *principle of symmetry*, implying that each fundamental observer sees himself to be at a centre of spherical symmetry.

As he later [1944] presented a proof demonstrating that the first principle, which is often interpreted as prescribing *universal homogeneity*, is contained in the second one, which entails *universal isotropy*, we here treat the two principles as if they were one. Since the principle of isotropy is the only cosmological principle relevant to the present paper, I feel free to ignore world-models incompatible with that principle such as, e.g., the rotating one of Gödel.

The principle of cosmic isotropy is a precondition for the definability of a cosmic time which is the most fundamental feature of the synthesis of ideas discussed in this book.

The first of Walker's results was a quadratic invariant implying a Riemannian geometry. The only difference to the general theory of relativity was its scope of greater generality with regard to the path descriptions of free particles: these were not necessarily given by geodesics. The technique used was that of continuous groups of transformations which, combined with the principle of isotropy, yielded a relation of congruence needed for the conservation of angles. The principles together implied the free paths to obey a variational principle of least action.

In this paper he deduced what became famous as the Robertson-Walker Metric (***RWM***):

$$(10) \qquad dT^2 = d\tau^2 - R^2(\tau)\left(\frac{d\rho^2}{1-\kappa\rho^2} + \rho^2(d\vartheta^2 + \sin^2\vartheta\, d\varphi^2)\right)$$

For $R(\tau) \equiv \tau$, $\kappa \equiv -1$, $\rho \equiv \sinh\alpha$, this is reducible to a metric of uniform expansion:

$$(11) \qquad dT^2 = d\tau^2 - \tau^2\left(d\alpha^2 + \sinh^2\alpha(d\vartheta^2 + \sin^2\vartheta\, d\varphi^2)\right)$$

Translating this into standard coordinates, using $c \equiv 1$, $t \equiv \tau\cosh\alpha$, $r \equiv \tau\sinh\alpha$, he obtained:

$$(12) \qquad dT^2 = dt^2 - dr^2 - r^2\left(d\vartheta^2 + \sin^2\vartheta\, d\varphi^2\right)$$

Here *kinematic frame time* t is connected with *private flat space*, *dynamic proper time* τ being connected with *public curved space*. Of these two time-scales, only τ is truly cosmic.

Metric (12), of course, is the quadratic invariance of the differential Lorentz equations. These equations are supposed to involve a reciprocal retardation of clocks in inertial motion.

But it is one thing to admit that the master-clock of a fundamental observer *may appear* retarded relative to the slave-clocks belonging to the frame of another fundamental observer. Quite another thing is it to claim that the master-clock of one fundamental observer *is in fact* retarded relative to the master-clock of another fundamental observer.

There is no problem in the first case due to the fact that slave-clocks are not fundamental. However, there *is* a problem with *the second case*, since it *involves a formal contradiction*! My point is that, if the master-clocks of two fundamental observers do not agree, they are not congruent, therefore at least one of the two fundamental observers cannot be fundamental.

That I am right, and a very widespread interpretation of relativity is wrong, is stressed by the fact that the RWM clearly not merely involves, but presupposes, a Cosmic Time!

That this is indeed the case is evidenced by Walker's remarks in §3 of his first paper:

"*Each observer can (thus) provide himself with a clock and can choose a coordinate system so that, for any two observers, their clocks are (congruent) .. in the sense (of) Milne. .. It may be assumed, therefore, that (all) clocks are .. synchronized (to) record the time τ .. Since there is a four-fold infinity of events, we may set up a (1,1) correspondence between events and points of a four-dimensional manifold M_4. Each fundamental particle occupies a single infinity of events during its history and corresponds to a world-line in M_4.*

*There is a world-line corresponding to each particle, and, since each event is occupied by one particle, there is just one fundamental world-line passing through each point of M_4. The points of each world-line may be specified by the times recorded by the clock associated with the corresponding particle, hence τ may be regarded as a parameter for each world-line. This parameter defines a family of three-dimensional surfaces S_τ ... Thus τ, which may be regarded as a **cosmic time**, serves as a (public) coordinate of M_4.*"

Later in the same paper (§16) he made these remarks, at once illuminating and intriguing: "*In §3 the clock recording time τ was any clock constructed by one observer. However, we already possess a clock recording time t which we may call "physical" (atomic? MTW) time, and all our time measurements are made with this clock. It is therefore desirable to refer to time t and, since the clock recording time τ was arbitrary, we may now assume that $\tau \equiv t$.*"

In that same paper Walker also verified Milne's expression for the density-distribution of a substratum and his equation for the motion of test-particles in the substratum as well as that for a statistical ensemble superposed hereon, both relating to Milne's uniform expansion model.

In three later papers on relativistic mechanics [1940 ff.], also inspired by Milne's ideas, he studied their relevance to a range of models comparable to the scope of general relativity.

According to Walker, "*the great disadvantage of general relativity*" is that it allows the form of space to depend on the distribution of matter, albeit the problem of finding that form is too difficult except for the simplest systems. A more serious objection is that no system can exist in isolation from the universe, so the problem of world-structure should be given primacy. The outstanding merit of Milne is, precisely, to have called attention to this crucial point.

Again Walker took *the principle of isotropy* as his starting point, this time setting himself the task of framing a general scheme differing from that of Milne by stressing even more the correspondence with classical mechanics. Seeing that the position and motion of an accidental particle is describable by reference to two fundamental ones: that which it is just passing and that relative to which it is now at rest, he showed that systems similar to those of classical mechanics are definable relative to the comoving frames of all fundamental observers.

He finally proved the theorem that all physical objects are describable by 3-space tensors transforming between fundamental particles. His general conclusion was this one: for each of the two time-scales there is a 3-space representation analogous to that of classical mechanics, and in both cases this 3-space representation is simpler than the 4-space representation.

Mogens True Wegener

4.   NON-STANDARD COORDINATES OF TIME

In other words there are universally acknowledged to be in heaven and earth things not dreamt of in Einstein's philosophy; yet Einstein's philosophy is allowed to block the simplest ways of categorizing and describing those things ... Thus it is taken for granted that quantum mechanics has to be "relativized", i.e., changed to conform to the Lorentz group, whereas it might more rationally be assumed that relativity theory and its invariance group have to be changed to conform to the existence in nature of nonlocal actions ...

T.E. Phipps: *Heretical Verities* .. [1986]

It is a dogma of the special theory of relativity (***SR***) that clocks are retarded and that rods are contracted, both as an effect of inertial motion. What are we to think of that dogma? Laws of nature being distinguished by invariance, how can the dilatation of time ever be a law? What reasons are adduced to substantiate it? Primarily a reference to the Lorentz-formulae!

Well, the Lorentz transformations are inevitable if we stick to the standard definition of coordinates proposed by Einstein and repeated ever since. But definitions are conventions. How can such conventions be inevitable? We are then referred to the so-called light-principle: If the velocity of light is a universal constant, how can the standard definitions be avoided?

A simple straigth-forward answer to this question is: by a little intuition and imagination! Further, how can a *state* of motion, supposed to be unaltered, ever be the *cause* of anything? Moreover, contractions are assumed to be reciprocal, retardations not; why this difference? How - disregarding bifurcations - can asymmetry ever be a consequence of symmetry?

In order to evade this impasse and make progress possible, a distinction must first be introduced, viz., that between the *one-way velocity* of light and the *two-way velocity* of light (to which we may add a generalization of the second concept: the round-trip velocity of light).

Today it is generally agreed that ***SR*** offers no experimental way of deciding the value of the one-way velocity of light which does not itself depend on a conventional definition of simultaneity-at-a-distance, tacitly taking the one-way light-velocity to be a universal constant. However, the procedures devised to test the two-way (or round-trip) velocity of light do not involve such definitional circularity. Hence the light-principle should be interpreted as referring not to the one-way light-velocity, but to the two-way (or round-trip) velocity of light instead.

This insight opens up for possibilities ignored by almost all experts on relativity theory, although the general theory of relativity (***GR***) similarly introduces a variable velocity of light; the problem here is that Riemannian geometry and tensors are assumed to be indispensable. But if the one-way velocity of light is finally recognized as a theoretical variable, the search for a rational alternative to the Einsteinian time-coordinate needs no longer be blocked.

The second point to be realized is that the *substratum* functions as a *compass of inertia*. The term, invented by Weyl, was used by Gödel to denote a reality distinct from the substratum, in order to make sense of a model where the substratum "rotates" relative to "something else"; but this empty conception is clearly unsatisfactory from the point of view of epistemology. The only rational procedure is to identify the two, and the natural precondition of doing so is that no spatial direction is privileged, as perceived by observers belonging to the substratum.

Thus, if two of its particles recede from each other with velocities depending on distance the whole substratum must expand in all directions according to the same function of time. Naturally, this can hold only if the same temporal parameter holds for the entire universe.

Non-Standard Relativity

An ideal substratum characterized by perfect symmetry can be treated as a cosmic "fluid" subject to the laws of hydrodynamics. However, our universe is not continuous, but discrete. How great are the irregularities it can absorb, and yet still retain an approximate symmetry? This is a difficult question allowing no easy answer, and we shall put it aside at present.

Let us instead agree to utilize *the idea of the substratum*, subject to perfect continuity, as a kinematic background - a neutral "sheet" of tempo-spatial geometry - on which we can plot irregularities in order to disclose the hidden pattern, or structure, of the observable universe, i.e., let us think of the substratum in the same way as the geometers think of coordinate systems. When interpreted thus, a substratum S is nothing but a reference frame in uniform expansion.

The *CP*, or the principle of *cosmic isotropy*, implies motion in a substratum to be radial: this holds for the relative motions of fundamental particles, whereas accidental particles deviate. Now, clearly, the comoving rest-frames of either particles have no real function of their own; this indicates the standard time-coordinates referring to such rest-frames to be redundant.

So we are free to start our search for alternatives to the standard conventions of Einstein. As mentioned above, experts today generally agree that the one-way light-speed (1-*wls*) may vary in accordance with the particular definition of simultaneity adopted as long as the average, or two-way, light-speed (2-*wls*) remains a universal constant. Writing $c_\to$ and $c_\gets$ for the 1-*wls* to and fro (out and home), respectively, the 2-*wls* c_o must fulfil the Light Principle *(LP)*:

$$\tfrac{1}{c_\gets} \equiv 1+\zeta \ . \ \tfrac{1}{c_\to} \equiv 1-\zeta \ . \ \tfrac{1}{2}\left(\tfrac{1}{c_\gets}+\tfrac{1}{c_\to}\right) \equiv \tfrac{1}{c_o} \equiv 1$$

This suggest a redefinition of Einstein's coordinates, $t \equiv \tfrac{1}{2}(\tau_3+\tau_1) , r \equiv \tfrac{1}{2}(\tau_3-\tau_1)$:

(13a) $\qquad\qquad t + r = \tau_3 \equiv \tau + r/c_\gets = (\tau+r\zeta) + r$

(13b) $\qquad\qquad t - r = \tau_1 \equiv \tau - r/c_\to = (\tau+r\zeta) - r$

In fact, at this stage we might even obtain $\tau = \tau_3$ by choosing $c_\gets \simeq \infty$ together with $c_\to = \tfrac{1}{2}c$. So, by taking the 1-*wls home* to be infinite, the idea of *quantum instantaneity* gets a new import. Inserting instead the invariant time $\tau \equiv \sqrt{t^2-r^2}$ of Törnebohm [1963], we find the deviation ζ:

(14) $\qquad\qquad \zeta = (t-\tau)/r = \{t-\sqrt{t^2-r^2}\}/r$

$$v \to r/t \ . \Rightarrow : \ \zeta \to \{1-\sqrt{1-v^2}\}/v \ \Rightarrow \ \tau \to t/\sqrt{1-v^2} \equiv t\gamma$$

This opens up for a radical re-interpretation of *SR*, call it *AR*, based on the idea that the deviation of 1-*wls* from 2-*wls* discloses a real - but maybe unobservable - difference: $SR \neq AR$. Accepting the Lorentz Transf.s for fundamental observers F & F', we obtain, for $v \equiv tanh\omega$:

$$x' = x\,cosh\omega - t\,sinh\omega \ . \ y' = y \ . \ z' = z \ . \ t' = t\,cosh\omega - x\,sinh\omega$$
$$x = 0 \wedge t = \tau . \Rightarrow x' = -\tau\,sinh\omega \ . \ x' = 0 \wedge t' = \tau . \Rightarrow x = \tau\,sinh\omega$$

Proper velocity has no upper limit. So why not put the master-clocks of F & F' on a par? This we do by relating fundamental observers pairwise *via* the frame of their *midway particle*. Applying the non-standard definitions $t^o \equiv t-x\,tanh\tfrac{\omega}{2}$, $x^o \equiv x-t\,tanh\tfrac{\omega}{2}$ to x' & t', we get:

$$x' = x-t^o sinh\omega \ . \ y' = y \ . \ z' = z \ . \ t' = t-x^o sinh\omega$$

The coordinates t^o & x^o refer to the comoving frame of a particle M midway between F & F'. So, for the midway particle, $x^o = 0$, whence: $t' = t$ & $x' = x = t\,tanh\tfrac{\omega}{2}$.

Ignoring x^o and the frame times t & t', we focus on $t^o \equiv t - x\,tanh\frac{\omega}{2}$ which we interpret as the common proper time τ displayed by the master-clocks of observers F, F', and that of M:

(15) $$x' = x\,cosh\omega - t\,sinh\omega = x - \tau\,sinh\omega$$
(16) $$\tau = t - x\,tanh\tfrac{1}{2}\omega = t' - x'tanh\tfrac{1}{2}\omega' = \tau'$$

Evidently, the *midway particles* between pairs of observers belonging to a substratum S are themselves members of that substratum, and the time shown by the master-clocks of all substratum members is invariant. This can be generalized to objects moving in the substratum. *All coordinates being communicated among members of S, any pair of observers $F, F' \in S$ may describe an object O by reference to substratum time τ, disregarding the frame times t, t'.* The observers F, F' are now able to determine the motional state of O in the following way:

While O's distance relative to F, resp. F', at instant τ is equal to that of particle $P_1 \in S$ with which O momentarily coincides, O's velocity relative to F, resp. F', at the same instant τ, is equal to that of particle $P_2 \in S$ with respect to which O is momentarily at rest, implying:

$$\vec{r}_{FO}(\tau) = \vec{r}_{FP_1}(\tau) \quad . \quad \dot{\vec{r}}_{FO}(\tau) = \dot{\vec{r}}_{FP_2}(\tau)$$
$$\vec{r}_{F'O}(\tau) = \vec{r}_{F'P_1}(\tau) \quad . \quad \dot{\vec{r}}_{F'O}(\tau) = \dot{\vec{r}}_{F'P_2}(\tau)$$

Now, from Walker's metric for the Milne universe it nevertheless follows that:

$$dT = \{d\tau^2 - \tau^2 d\sigma^2\}^{\frac{1}{2}} = \{dt^2 - dr^2\}^{\frac{1}{2}} = invar.$$

For motion relative to the substratum ($d\sigma \neq 0$), **AR** still predicts a retardation of clocks. Such *clock retardation*, however, is not just an effect of inertial motion, as generally assumed. In Milne's universe, it can always be interpreted as *a gravitational effect* of the substratum S.

In any case, $d\sigma \to 0 \Rightarrow d\tau \to dT$ shows that the deviation of τ from true time T depends on the deviation of the observed object from fundamentality, i.e., on its motional anisotropy.

5. TOWARDS A NEW STEADY STATE THEORY

In any theory which contemplates a universe (evolving with time), explicit and implicit assumptions must be made about the interactions between distant matter and local laws .... These are necessarily of a highly arbitrary nature, and progress on such a basis can only be indefinite and uncertain. It may .... be questioned whether such speculation is required. If the universe (is in a steady state) none of these difficulties arises.

Sir Hermann Bondi [1961³]

The idea that the universe originated in a dramatic explosion, the "big bang", appears to be supported by two important discoveries: (i) that of a systematic redshift in the light from distant galaxies [Hubble 1929], and (ii) that of a faint cosmic background radiation displaying a highly isotropic character [Penzias & Wilson, and Dicke, 1965]. It has since become one of the most cherished dogmas of modern science.

But the redshift in itself does not imply a singularity at $t = 0$. Further, other explanations of the 3K **CBR** than a "big bang" are open, cf. Surdin [2002]. Narlikar [1980] reckons several alternatives. Finally, the "big bang" idea has recently run into severe difficulties. As pointed out by Lerner [1991], certain huge galactic structures seem to be as old as a hundred billion years, and in later years new reports of very distant, but fully mature, galaxies abound.

During the seventies, the "big bang" hypothesis gradually superseded the once popular "steady state" theory of Bondi, Gold & Hoyle; cf. North [1965], Harrison [1981], Kragh [1996]. However, it would be wrong to suppose that the "big bang" hypothesis and the "steady state" hypothesis are mutually exclusive. In fact, it is easy to construct world-models which originate as a point-singularity, or even as a quantum-state of infinite density, and which then later by a temporal evolution, fierce or gentle, progressively approximate towards a stationary state.

If approximation is allowed to take place, at least two interesting possibilities are open: (1) a universe which starts as a singularity and expands in hyperbolic space according to a hyperbolic sine function of time, and (2) a universe which begins like "a cosmic egg" and then develops in hyperbolic space in agreement with a hyperbolic cosine function of time.

In his first book, Milne proposed a distinction between two important concepts relating to any rational model of the universe: (1) its *world-map*, describing the universe as a totality of structure that exists independently of observation; and (2) its *world-view*, depicting the universe in perspective as perceived by an observer. Whereas the world-map is a structural "snapshot" presupposing a definition of universal simultaneity, the world-view combines spherical "layers" of varying age, their age increasing with their radial distance from the observer, due to the fact that we usually take the one-way velocity of light to be finite. A genuine "steady state" model is characterized by a sort of identity between the two, whereas for "big bang" models they differ. Most cosmologists ignore the difference, being unable to make sense of the concept of world-map, but this invalidates their interpretation of data. Indeed, one could say that their refusal to transcend a pure world-view excludes them from developing cosmology as a theory.

Let us consider the original model of Bondi & Gold which is based on the metric:

$$(17) \qquad dT^2 = dt^2 - t_o^2 \, e^{2t/t_o} \left(d\rho^2 + \rho^2 (d\vartheta^2 + sin^2\vartheta \, d\varphi^2) \right)$$

This simple model with comoving coordinate ρ and cosmic time t is invariant to time-zero shift; but other properties are less agreeable: $dT = d\vartheta = d\varphi = 0$ yields $\pm \, d\rho = e^{-t/t_o} dt/t_o$, whence:

$$(18) \qquad \rho \underset{t_2 = 0}{\equiv} \int_{t_1}^{0} e^{-t/t_o} dt = \int_{0}^{t_3} e^{-t/t_o} dt = const.$$

$$(19) \qquad \rho = 1 - e^{-t_1/t_o} = e^{-t_3/t_o} - 1$$

For reflection at $t_2 = 0$, (19) shows that: (a) $e^{t/t_o} \rho$ has a horizon at $t_3 \to \infty \Leftrightarrow t_1 \to -t_o \, ln2$; (b) all particles receding from a finite distance leave our part of the universe after a finite time; (c) only a finite number of particles are at any time visible within the horizon, whereas an actual infinity of particles does exist in an imaginary space beyond the horizon. Now (c) reduces the visible universe to a finite drop in an infinite ocean of "reality". This is highly problematic.

There are *two main reasons to search for an alternative to this model* of the universe. The first reason is theoretical: it is gross speculation to assume almost the entire universe to exist beyond the horizon of all possible observation as a hardly intelligible kind of reality. The second reason is empirical: it has turned out to be hard to reconcile the observed number-distance statistics for galaxies with that predicted by the model; but cf. Hoyle & al.[2000].

In contrast to that, the uniform expansion model of Milne is unbounded by any horizon, and hence it contains *a potential infinity of visible objects* inside its imaginary circumference. It is therefore natural to ask whether these properties of Milne's "big bang" model could not be transferred to a new "steady state" model. We shall address this question in §6.

Mogens True Wegener

Until then, it seems mandatory that we discuss some important preliminary difficulties. Unfortunately, we are forced to leave the most important one unsolved, viz., that concerning the degree of tolerance against irregularities and deviations from perfect universal symmetry. But it is intuitively clear that, since our new model of a "steady state" universe should be constructed so as to leave no real (i.e., existing) light source invisible or unobservable in principle, it will also be able to accomodate much vaster irregularities and differences of development. Hence patterns as that of the "great wall" do not in themselves make the model improbable.

Another difficulty, related to the first one, is that the concept of a linear equivalence cannot be applied to a "steady state" model which must be invariant to arbitrary time-zero shift. We shall therefore have to define the equivalence class of observers in a manner that makes it independent of the form of signal-functions and their commutativity for collinear observers. This we do by postulating clock-congruence to be transitive for all fundamental observers:

$$\forall P, Q, R: \ \boldsymbol{T}^p \simeq \boldsymbol{T}^q \ \& \ \boldsymbol{T}^q \simeq \boldsymbol{T}^r \ . \Rightarrow \ \boldsymbol{T}^p \simeq \boldsymbol{T}^r$$

So we define a substratum in this new sense as an infinite set of fundamental particles that are equivalent by virtue of the fact that their master-clocks remain mutually congruent. Our definition corresponds exactly to a tacit assumption of orthodox cosmology: namely, that *atoms of the same type oscillate with the same frequency when unaffected by local forces.* This assumption is clearly indispensable to all natural science, especially cosmology. Our only addition to it is that such atoms rule the clocks of a universal class of equivalent observers.

The advantage of this interpretation is that it changes the status of fundamentality from something absolute into a matter of degree: the less the atoms regulating the master-clock of an observer-particle deviate from the ideal standard of cosmic time, the more fundamental it is. By the same token,, the deviation of a master-clock from universal time becomes a measure, or an index, of the influence upon it caused by local gravity, or other kinds of force.

Hence cosmic time is an ideal, just like the perfect vacuum, the perfect inertial motion, or the absolute zero of temperature; but natural science, of course, is in need of such ideals. Bondi sees it as a merit of his own "perfect cosmological principle" that it entails a maximum of uniformity in nature; only on this condition, says he, can nature be subject to strict laws. But, as already hinted at, it may be doubted whether nature is really governed by strict laws.

Probably it should be reckoned as a merit of a world-model if it allows a maximum of non-uniformity in nature without surrendering its structural totality and fundamental integrity. Let it be admitted that *cosmic time*, as represented by *atomic clocks*, is a statistical concept; but, as an index for deviation from universality it is undoubtedly also a crucial regulative idea, and *it is a legitimate task to construct world-models that conform to this regulative idea.*

The "steady state" model of Bondi, Gold and Hoyle, implying the continued creation of "matter" in "space", has often been the target of heavy criticism for its presumed violation of one of the two most imperative principles in science, viz., that of the conservation of energy. This vital principle of physics is frequently assumed to be derivable from the ancient maxim *ex nihilo nihil fit* without recognition of the fact that this maxim is a piece of metaphysics.

Moreover, the participants in this often emotional debate repeatedly forget that, in order to state the energy principle properly, mention must be made of a specific volume of space wherein a certain amount of energy is supposed to exist. How is this volume to be defined? How can a volume be specified without presupposing a definition of distant simultaneity?

Non-Standard Relativity

Let us yet suppose that a certain mass is definable as the product of density by volume, and that a volume is definable by its spatial extension as measured by radar signals; even with the assumption of an absolute simultaneity as a cut "at right angles" to the universal time axis, the question as to the time-scale to be used is also ambiguous: what time should be counted? Let us then settle for the scale of atomic time, and still one question is left for us to decide: do we want to measure spatial volume by means of proper distance or coordinate distance?

If the question is stated that precise, the problem of energy becomes one of hypothesis, and the difference between "big bang" models and "steady state" models can be specified thus: *In a pure "big bang" model energy conservation is defined relative to constant coordinate space whereas, in a pure "steady state" model, it is defined with respect to constant proper space.* But it is hard to see why coordinate space should be given priority over proper space.

In fact, proper distance defined as radar-distance with respect to cosmic time appears to be closer to our ordinary intuition of "reality" than any convention of coordinate distance. The only conundrum to our intuition is that energy conservation as defined for an expanding substratum relative to proper space leads to the consequence that the apparent loss of matter caused by the recession of particles has to be compensated by the creation of other particles.

This puzzle finds a solution when it is realized that the model of continued creation here presented allows no particle to cross its imaginary border: there is nothing beyond the horizon. Such universe can be described as *a pseudosphere in flat space*, of finite radius and finite mass, containing an infinite, yet apparently ever growing, number of atoms which apparently shrink with their distance from an observer, reaching zero value precisely as they pass its periphery. *Neither light nor matter ever leaves the universe that thus simulates a perfect "black hole".*

Moreover it should be realized that "big bang" models and "steady state" models have a different status with respect to the presumed "constants" of nature. While "big bang" models are open to the possibility of a secular variation of their values and their numerical proportions, the invariant structure of "steady state" models urges the need for a theoretical explanation. The idea of Eddington that a theory of cosmic numbers may be feasible is indeed fascinating; cf. Whittaker [1943], who for that reason described Eddington as "a modern Archimedes".

Finally, according to Milne's theory, gravity does not hamper the universal expansion. So what could be the mechanism explaining the acceleration needed in a "steady state" model? In any model of non-zero density there is a certain pressure that might cause an expansion. If the stuff of the universe can be viewed as a gas, the internal pressure of this gas might explain universal dissipation, hence also gravity, leaving the universe itself as the only "black hole".

6. A PROGRAM FOR SYNTHESIS

In physics everything depends on the insight with which the ideas are handled before they reach the mathematical stage.   A.S. Eddington [1939 ch.4].

The program of *Einstein* aimed at *a geometrization of physics*, as he often repeated. That of *Milne*, by contrast, aimed at *an arithmetization of physics*, to use a phrase of Whitrow. These characterizations reflect the fact that, whereas Einstein buried the classical forces in the geometric structure of space - indeed a strange bearer of properties! - Milne set them free by deriving them as the arithmetic result of local perturbances in the global expansion.

Mogens True Wegener

Gravity in his theory does not act as a brake on the expansion which remains uniform. Thus, in a way, Milne confirms the famous conjecture of the atomists: viz., that real objects, as well as all forces operating on them, stem from the "free fall" of atoms through empty space. Another characterization of that part of their programs, due to myself (1990), runs as follows: *Whereas the aim of **Einstein** was to reduce **inertia to gravitation**, in accordance with Mach's principle, that of **Milne** was to reduce **gravitation to inertia**, in opposition to Mach's principle.* Only the latter procedure can free physics of one of its greatest mysteries: that of gravity.

From a philosophical point of view there is no doubt, to my mind, that Milne has offered not only the best but, indeed, the only reasonable explanation of the mechanism of gravitation. What he has presented us with is a world-model which allows a derivation of all so-called "laws of nature" from first principles by superposing layers of accidental particles upon the universal substratum of fundamental particles supposed to be in uniform dispersion.

This ***kinematic technique*** was since generalized by Walker who succesfully applied it to a whole range of world-models (universes) with different expansion-functions and geometries, and who demonstrated the value of ***multiple time-recording***, though neither the kinematic scale nor the dynamic scale should of necessity be associated with the intrinsic rhythm of atoms. The point is that the substratum is represented as being stationary according to one scale and uniformly expanding according to the other; but this says nothing about atomic rhythms.

Now, if the kinematic technique is to be applied at all to a "steady state" universe of continued creation, that model must display some analogy to the ideas of Milne and Walker. The standard metric of the universe of Milne, as well as that of Törnebohm & Prokhovnik, is:

$$dT^2 = dt^2 - dr^2 - r^2(d\vartheta^2 + sin^2\vartheta \, d\varphi^2)$$

I now suggest that we take over the same metric, only adapting it to hyperbolic 3-space,

(20) $$dT^2 = dt^2 - dr^2 - sinh^2 r \, (d\vartheta^2 + sin^2\vartheta \, d\varphi^2)$$

to characterise a new world-model, or universe, representing the continued creation of matter. The diffusion of fundamental particles in this universe is described by the following differential equations in stead of the linear equations $r = t \, tanh\alpha = \tau \, sinh\alpha$ characterizing Milne's model:

$$v \equiv dr/dt \equiv tanh(r/r_o) \propto r \text{ for } r \ll r_o$$
$$w \equiv dr/d\tau \equiv sinh(r/r_o) \propto r \text{ for } r \ll r_o$$

These equations immediately yield the following fundamental expressions so distinctive of **SR**:

(21) $$\gamma_r \equiv dt/d\tau = cosh(r/r_o) = \sqrt{1+w^2} = 1/\sqrt{1-v^2}$$

Following Milne, the Doppler-Shift (*DS*) is the derivative of the signalfunction θ, cf. (7):

(22) $$s(t) = 1+z(t) = \{(dt+dr)/(dt-dr)\}^{1/2} = e^{r(t)}$$

A natural choice of distance-unit r_o is where the *DS* assumes the value e:

(23) $$s(t) = 1+z(t) = e^{r/r_o} = e \Leftrightarrow r = r_o = t_o \equiv 1$$

According to Prokhovnik [1967, §6+app.] the expression $1+z = e^r$ matches the observed number-redshift relation which was utilized by astronomers to disprove the "steady state" model of Bondi & al.; but he ignores the structurak difference between *world-map* and *world-view*, redundant to a "steady state" model, but essential to his "big bang" model, identical to Milne's. So, rather than supporting the Milne-Prokhovnik model, it supports the one proposed here.

=//=

Non-Standard Relativity

CHAPTER 4

VARIOUS COSMOLOGICAL MODELS
THEIR TIME SCALES AND SPACE METRICS

=//=

*Revised version (2015,2020,2021) of a paper
presented at PIRT 8, Imperial College, London, 2002*

=//=

SUMMARY

Provisionally accepting the standard Robertson-Walker Metric (RWM) of cosmology and recalling that the principle of cosmic isotropy can be used as an argument for the definability of an all-embracing universal time, at least statistically, we propose to reverse this procedure by postulating such time as a regulative idea in the sense of Kant.

=//=

CONTENTS

=//=

Mogens True Wegener

1. INTRODUCTION

According to Milne & Walker, two kinds of observer-particles should be distinguished: *fundamental ones*, defining the structure of the specific cosmological model under consideration by *constituting its substratum*, and *accidental ones* which are *superposed upon the substratum*. The substratum thus serves as a universal "rest-frame", or "compass of inertia" (Weyl).

Taking the Robertson-Walker Metric (*RWM*) as our point of departure, we investigate the properties of two cosmological models: *a*) the *uniform expansion* model of Milne, which is the simplest model of a cosmic "big bang", and *b*) the *exponential expansion* model of Bondi & Gold, generally supposed to be the simplest possible model of a cosmic "steady state".

In accordance with a tentative acceptance of *RWM* we consider the relationship between our choice of time scale for a certain model of the universe and its corresponding space metric. As it turns out, there are at least two important ways of mapping the expansion of the universe: *a*) one which keeps atomic sizes constant while light is being stretched, and *b*) one which keeps distances between fundamental observers constant while their atoms are shrinking.

Finally, a new "steady state" model of the universe is proposed that deviates from *RWM* by allowing atoms to be contracted due to universal dispersion. In this model, spatial curvature is apparently increasing with the distance at which an object is seen by a fundamental observer. Constructed directly from $d\mathcal{T}^2 = dt^2 - ds^2 = dt^2 \gamma^{-2}$ (*SR*), it is simpler than that of Bondi-Gold. The basic formal properties of this model and two related ones are then examined.

Added 2015:

As regards material properties it is obvious that, in a "steady state" universe, all constants of nature must be genuine constants relative to that time-scale which preserves the atomic radii. Therefore we cannot accept Milne's idea that Planck's constant varies relative to that scale.

By contrast, we are confronted with a choice as regards the validity of the law of inertia: does that law hold relative to the scale which preserves the atomic radii, or relative to that which preserves the distances between fundamental particles, thus leaving the substratum static?

Here we follow Milne, assuming the law of inertia to hold relative to the latter scale only. As a consequence, inertial motion relative to the latter scale must be described as accelerated relative to the former scale, i.e., that scale which preserves the atomic radii invariant.

This seems to open up the possibility for an explanation of the rotation pattern of galaxies and their clusters which makes the introduction of socalled "dark matter" redundant.

Added 2020:

The chapter has been abbreviated and simplified as compared to older versions.

Added 2021:

Some further simplifications have been made concerning the sections §§5&6.

=//=

Non-Standard Relativity

2.   A CLASSICAL ALTERNATIVE TO SR

The hyperbolic formulae corresponding to the standard addition $\alpha' \equiv \alpha - \omega$ are:
$$cosh\alpha' = cosh\alpha\,cosh\omega - sinh\alpha\,sinh\omega$$
$$sinh\alpha' = sinh\alpha\,cosh\omega - cosh\alpha\,sinh\omega$$
Using $c_o \equiv unity$, $v \equiv tanh\omega$, the **LT** of **SR** with temporal coordinates become:
$$t' = t\,cosh\omega - x\,sinh\omega \ . \ \ x' = x\,cosh\omega - t\,sinh\omega$$
It is interesting that the *LT* are derivable from these addition formulae if and only if $\tau \equiv \tau'$ in:
$$t'/\tau' \equiv cosh\alpha' \ . \ \ t/\tau \equiv cosh\alpha \ . \ \ x'/\tau' \equiv sinh\alpha' \ . \ \ x/\tau \equiv sinh\alpha$$
It is natural to identify τ^2 with the **SR** invariant: $\mathcal{T}^2 \equiv t^2 - x^2 - y^2 - z^2 = t'^2 - x'^2 - y'^2 - z'^2$.

The standard (1x3) *LT* for three inertial frames, Σ, S & S', in relative motion are:

(1a) $\qquad X \equiv x\,cosh\alpha - t\,sinh\alpha = x'cosh\alpha' - t'sinh\alpha' \ . \ Y \equiv y \equiv y'$

(1b) $\qquad T \equiv t\,cosh\alpha - x\,sinh\alpha = t'cosh\alpha' - x'sinh\alpha' \ . \ Z \equiv z \equiv z'$

Consider Σ to be a preferred frame with the privileged observer Ω situated in its *origo*. Let the observers O & O' be situated in the *origos* of S & S', resp., and let the frame times t & t' of S & S' be synchronized to the proper time T of Ω by choosing $T \equiv t \equiv t' \equiv 0$ when O & O' both coincide with Ω. Suppose that an event E occurs at particle P, as observed by Ω, O & O'. Let the standard coordinates of E be (T, X, Y, Z) in Σ, (t, x, y, z) in S and (t', x', y', z') in S'. Then, by eliminating the irrelevant frame times t & t' from the expressions for T & X, we get:
$$X = x/cosh\alpha - T\,tanh\alpha = x'/cosh\alpha' - T\,tanh\alpha'$$
Further, using $\omega = \alpha - \alpha' = -\omega'$ to eliminate α or α', we recover **LT** for the privileged time T:

(2a) $\qquad x' = \{x\,cosh(\omega - \alpha) - T\,sinh\omega\}/cosh\alpha$

(2b) $\qquad x = \{x'cosh(\omega' - \alpha') - T\,sinh\omega'\}/cosh\alpha'$

Finally, introducing non-standard frame-times $\bar{\tau}$ & $\bar{\tau}'$ for frames S & S' defined by means of :

(3) $\qquad\qquad \bar{\tau} \equiv T/\gamma \ . \ \ \bar{\tau}' \equiv T/\gamma'$
$$\gamma \equiv cosh\alpha \equiv 1/\sqrt{1-v^2} \ , \ \ \gamma' \equiv cosh\alpha' \equiv 1/\sqrt{1-v'^2}$$

with $w \equiv tanh\omega$, we recover the Tangherlini transformations *(TT)* as generalized by Selleri:

(4a) $\qquad x' = x\,\dfrac{cosh(\omega - \alpha)}{cosh\alpha} - \bar{\tau}\,sinh\omega = \dfrac{x(1-wv)-w\bar{\tau}}{\sqrt{1-w^2}}$

(4b) $\qquad x = x'\,\dfrac{cosh(\omega' - \alpha')}{cosh\alpha'} - \bar{\tau}'sinh\omega' = \dfrac{x'(1-w'v')-w'\bar{\tau}'}{\sqrt{1-w'^2}}$

In standard *SR*, it is always the proper time of a single moving clock which is said to be retarded relative to the slave clocks distributed as a network over the rest frame of the observer; but if we refer the inertial motion of particles to a privileged frame we should use *TT* instead. Please notice that *TT* reduce to *GT* if all observations refer to the frame of the midway particle:

(5) $\qquad\qquad \alpha = \frac{\omega}{2} \ \Rightarrow \ x - x' = \bar{\tau}\,sinh\omega = 2T\,sinh\frac{\omega}{2}$

Applying $\bar{\tau} \equiv t - x\,tanh\frac{\omega}{2} \equiv t' - x'tanh\frac{\omega'}{2} \equiv \bar{\tau}'$ directly to *LT* we get the same result, viz., *GT*:

(6) $\qquad\quad \underline{\bar{\tau}' = \bar{\tau} \ , \ \omega' = -\omega \ , \ x' = x - \bar{\tau}\,sinh\omega \ , \ y' = y \ , \ z' = z}$

Mogens True Wegener

3. THE ROBERTSON-WALKER METRIC

In his classic monograph *Natural Philosophy of Time* (1961/1980), G.J. Whitrow devised a method to deduce the **RWM** of relativistic standard cosmology from the γ-factor of **SR**. Let:

$$c_o \equiv 1 \; . \; c_o t_o \equiv r_o \equiv 1$$

and let the *origo* of the comoving standard rest frame S of an observer P be P himself.

Now suppose an event E, taking place at some object O, to be triggered by a light signal which instantaneously released a visible flash. Suppose further that this light signal was sent off by P at the instant τ_1, and that the flash was received by P at the instant τ_3, both τ_1 & τ_3 being instants of proper time τ of P as read off his own standard atomic clock C. We then recover the Einstein coordinates of the Cartesian frame S of P by means of the usual definitions:

$$\tau_3 \equiv t+r \; . \; \tau_3' \equiv t'+r'$$
$$\tau_1 \equiv t-r \; . \; \tau_1' \equiv t'-r'$$

From the standard **SR** invariant $d\tau_3 d\tau_1$ we get the γ-factor for the retardation of moving clocks:

$$d\mathcal{T}^2 \equiv d\tau_3 d\tau_1 = dt^2 - dr^2 \equiv dt^2 \gamma^{-2}$$

Whitrow then suggested: $dr \rightarrow \mathcal{S}(T)\,d\sigma$. The parameter T of his function $\mathcal{S}(T)$ is thus no longer the *private frame time t*, but rather the *public proper time τ* of fundamental observers, i.e., all observers at rest in the universe, e.g., relative to the cosmic background radiation *(CBR)*. With $T \equiv \tau$, the standard invariant of **SR** is finally transformed into the standard **RWM** metric:

$$d\mathcal{T}^2 = dt^2 - dr^2 = d\tau^2 - \mathcal{S}^2(\tau)\,d\sigma^2$$

Here $\mathcal{S}$ is taken to be the *universal scale factor* and σ a fixed ("comoving") coordinate. Now, for *fundamental observers $d\sigma = 0$*, i.e., $d\mathcal{T} = d\tau$, showing that all fundamental observers are in pace with the same common *cosmic time $\mathcal{T}$*. By implication, any deviation of proper time τ from $\mathcal{T}$ is confined to non-fundamental, i.e., *accidental observers* discerned by a variable σ. Considering $\mathcal{T} \neq \tau \neq t$, one may ask if all this amounts to more than a vague analogy?

According to the standard view, it is always the *proper time* of a "moving" particle that is claimed to be "slow" relative to the *frame time* of a "stationary" observer. So coordinate time, or frame time, is thereby tacitly assumed to represent the "true extended time" of any observer. The cosmic time $\mathcal{T}$ implied by **RWM** is seldom taken seriously, but mostly ignored or explained away as being of "statistical origin" and thus "ill defined". Nevertheless, it is the firm stance of the present writer that a fundamental importance should be ascribed to the cosmic time $\mathcal{T}$.

If we define *true time* by the readings of the master clocks of our fundamental observers when they have been properly synchronized - e.g., by letting a definite non-local cosmic event, such as the beginning of everything in a socalled "big bang", represent a common time zero - then it is no longer true to say that the master clock of a fundamental observer is slow relative to the frame clocks of another observer, fundamental or not. Much rather it is true to say that it is the clocks of accidental particles that are slow relative to the clocks of fundamental observers. But the only conflict at stake here is one relating to the standard interpretation of **SR**.

Hence, if the clocks of fundamental observers show the true time $\mathcal{T}$, then the clock of an accidental particle will be more or less slow. In fact, *the greater its distance to that fundamental particle relative to which it is momentarily at rest, and which thus constitutes the natural origo of its own rest frame, the slower its clock will run* and the more it will deviate from true time. The natural way of interpreting this retardation of moving clocks is as an effect of gravitation.

In this way we have found a simple coupling between the rates of non-fundamental clocks and what seems to be a gravitational potential due to the substratum of fundamental particles.

The reason for this dependence is that *the deviation of the clock of an accidental particle from true time T is found by direct comparison with the clock of that fundamental particle with which it momentarily coincides*; and the greater the distance of an accidental particle is to the origo of that rest frame to which it belongs, the faster its speed relative to that fundamental particle with which it coincides will appear; this follows from the expansion function $\mathcal{S}(\tau)$. What we have disclosed is the possibility of an influence of the substratum on particles which do not belong to the substratum and which represent deviations from cosmic symmetry.

This supports the stance of Whitrow's former master, E.A. Milne. The essential point of his **Kinematic Relativity (KR)**, devised as an alternative to Einstein's theories, **SR & GR**, is that what we call gravitational effects may be due to local deviations from cosmic symmetry. Indeed, if elevated to a universal principle, *Milne's conjecture amounts to nothing less than an inversion of Mach's principle: while Mach held that inertia should be reduced to gravitation, Milne insisted that gravitation should be reduced to inertia* - and showed how to do it!

Now, with polar coordinates, the **RWM** can be written:

(6) $$dT^2 = d\tau^2 - \mathcal{S}^2(\tau)\{d\rho^2 + \lambda^2(d\theta^2 + \sin^2\theta\, d\phi^2)\}$$

(7) $$\mathcal{R} \equiv \mathcal{S}(\tau)\rho \,,\ \ T \equiv \int d\tau/\mathcal{S}(\tau) + const.\,,\ \ \mathcal{C}(T) \equiv dT/d\tau \equiv 1/\mathcal{S}(\tau)$$

Here $\mathcal{R}$ is proper distance, $\mathcal{C}$ is an inverse scale function, and T is an auxiliary time parameter.

$$d\rho \equiv d\lambda/\sqrt{1-\kappa\lambda^2} = \begin{cases} d\lambda & \Leftarrow \kappa = 0 \\ d\arcsin\lambda & \Leftarrow \kappa = 1 \\ d\,arsinh\lambda & \Leftarrow \kappa = -1 \end{cases}$$

$$dT^2 \underset{\kappa=0}{=} d\tau^2 - \mathcal{S}^2(\tau)\{d\rho^2 + \rho^2(d\theta^2 + \sin^2\theta\, d\phi^2)\}$$
$$dT^2 \underset{\kappa=1}{=} d\tau^2 - \mathcal{S}^2(\tau)\{d\rho^2 + \sin^2\rho(d\theta^2 + \sin^2\theta\, d\phi^2)\}$$
$$dT^2 \underset{\kappa=-1}{=} d\tau^2 - \mathcal{S}^2(\tau)\{d\rho^2 + \sinh^2\rho(d\theta^2 + \sin^2\theta\, d\phi^2)\}$$

$$d\rho \equiv d\lambda/\sqrt{1-\kappa\lambda^2} \equiv d\varrho/(1+\kappa\varrho^2/4) = \begin{cases} d\varrho & \Leftarrow \kappa = 0 \\ d\arctan(\varrho/2) & \Leftarrow \kappa = 1 \\ d\,artanh(\varrho/2) & \Leftarrow \kappa = -1 \end{cases}$$

$$d\rho^2 + \lambda^2(d\theta^2 + \sin^2\theta\, d\phi^2) \equiv \frac{d\varrho^2 + \varrho^2(d\theta^2 + \sin^2\theta\, d\phi^2)}{1+\kappa\varrho^2/4} \equiv \frac{d\xi^2 + d\eta^2 + d\zeta^2}{1+\kappa\varrho^2/4}$$

The expressions below cover all possible values of the constant of curvature, κ:

(8a) $$dT^2 = d\tau^2 - \mathcal{S}^2(\tau)\{d\lambda^2/(1-\kappa\lambda^2) + \lambda^2(d\theta^2 + \sin^2\theta\, d\phi^2)\}$$

(9b) $$= d\tau^2 - \mathcal{S}^2(\tau)\{d\varrho^2 + \varrho^2(d\theta^2 + \sin^2\theta\, d\phi^2)\}/(1+\kappa\tfrac{\varrho^2}{4})$$

(9c) $$= \mathcal{C}^{-2}(T)[\,dT^2 - \{d\xi^2 + d\eta^2 + d\zeta^2\}/(1+\kappa\tfrac{\varrho^2}{4})\,]$$

Applying the T-scale, the *expansion of cosmos* is explained away as a *shrinking of atoms*! In standard presentations, great importance is generally ascribed to the Hubble function $\mathcal{H}$:

$$\mathcal{H}(\tau) \equiv \dot{\mathcal{S}}(\tau)/\mathcal{S}(\tau)$$

Mogens True Wegener

4. MILNE'S SIMPLE BIG BANG MODEL

In what follows we throw light on the **RWM** by discussing some simple world models. One of the simplest is Milne's model of uniform expansion, adopted by Prokhovnik [1967]:

$$(9) \qquad \mathcal{S}(\tau) \equiv \tau/\tau_o \equiv d\tau/dT \equiv e^{(T-\tau_o)/\tau_o} \equiv \mathcal{C}^{-1}(T)$$

$$\mathcal{H}(\tau) \equiv \dot{\mathcal{S}}(\tau)/\mathcal{S}(\tau) \propto 1/\tau$$

Let us assume that radar signals are being exchanged between a pair of observers, $P \& Q$, in $(1\text{x}1)$ *timespace*. Suppose that a "photon" ϕ is emitted from P at $\tau = \tau_1^p$, reflected by Q at $\tau = \tau_2^q$, and received by P at $\tau = \tau_3^p$. Then, according to the relativity principle as interpreted by Milne, τ_3^p is the same function of τ_2^q as τ_2^q is of τ_1^p - call it $s(\tau) \equiv e^{\sigma}\tau$. Generalizing, and introducing Einsteinian standard coordinates $t \& r$ for P (priming those of Q), we at once get:

$$t \equiv \tfrac{1}{2}(\tau_3+\tau_1) \quad \Downarrow \quad r \equiv \tfrac{1}{2}(\tau_3-\tau_1)$$
$$\tau_3 = \tau\, e^{\sigma} = t + r \ . \ \tau_1 = \tau\, e^{-\sigma} = t - r$$
$$(10) \qquad t = \tau\, cosh\sigma \ . \ r = \tau\, sinh\sigma$$

Let us next assume that σ is not a constant, but a variable; then, by differentiation:

$$dt = d\tau\, cosh\sigma + \tau\, d\sigma\, sinh\sigma$$
$$dr = d\tau\, sinh\sigma + \tau\, d\sigma\, cosh\sigma$$

$$d\mathcal{T}^2 \equiv dt^2 - dr^2 = d\tau^2 - \tau^2 d\sigma^2 = e^{2(T-\tau_o)/\tau_o}(dT^2 - d\sigma^2)$$

This invariant is easily expanded into a hyperbolic *timespace* of $(1\text{x}3)$ dimensions if we put:

$$d\sigma^2 \equiv d\rho^2 + sinh^2\rho\,(d\theta^2+sin^2\theta\, d\phi^2) \equiv \{d\xi^2+ d\eta^2+ d\zeta^2\}/(1-\tfrac{\varrho^2}{4})$$

The standard **SR** invariant is thus transmuted into the hyperbolic metric of an expanding universe with expansion function $\mathcal{S}(\tau) \equiv \tau$, which can be transformed into another hyperbolic metric, viz., that of a stationary universe whose atoms all contract in agreement with the Hubble function $\mathcal{C}^{-1}(T) \equiv e^{(T-\tau_o)/\tau_o}$, where $T = \tau_o\{1+ log(\tau/\tau_o)\}$, τ_o being a constant of calibration:

$$(11a) \qquad d\mathcal{T}^2 = dt^2 - dr^2 - r^2(d\theta^2+sin^2\theta\, d\phi^2)$$

$$(12b) \qquad = d\tau^2 - \tau^2\{d\rho^2+ sinh^2\rho\,(d\theta^2+sin^2\theta\, d\phi^2)\}$$

$$(12c) \qquad \underset{\tau_o \equiv 1}{=} e^{2(T-1)}[\, dT^2-\{d\xi^2+ d\eta^2+ d\zeta^2\}/(1-\tfrac{\varrho^2}{4})\,]$$

The **1**st of these metrics, incorporating the universal constancy of the velocity of light, yields an infinity of *private timespaces*, comprising the flat 3-spaces of fundamental observers. The following two metrics both yield *a public timespace*, each containing a curved 3-space: that of the **2**nd metric being associated with τ-*time*, relative to which atoms keep constant sizes while the distances between fundamental observers steadily expand in proportion to $\mathcal{S} \equiv \tau$, with the consequence that *light is stretched*, as propsed by Prokhovnik - and that of the **3**rd metric being associated with T-*time*, relative to which distances between fundamental observers remain invariant whereas the sizes of their atomic constituents are contracting in proportion to $\mathcal{C}^{-1} \underset{\tau_o=1}{=} e^{T-1}$, due to *a secular reduction* of the velocity of light, as explained by Whitrow.

Non-Standard Relativity

5.   THE FIRST STEADY STATE MODEL

Now, passing from Milne's world model to that of Gold & Bondi, and of Hoyle, the scale factor $\mathcal{S}(\tau)$ is changed from $\tau \equiv \tau/\tau_o$ to $e^\tau \equiv e^{\tau/\tau_o}$, characterizing the "steady state" model: so:

(12)
$$\mathcal{S}(\tau) \equiv e^\tau \equiv d\tau/d\mathrm{T} \equiv \tfrac{1}{1-\mathrm{T}} \equiv \mathcal{C}^{-1}(\mathrm{T})$$
$$\overline{\mathcal{H}(\tau) \equiv \dot{\mathcal{S}}(\tau)/\mathcal{S}(\tau) \propto const.}$$

$\mathcal{R} \equiv e^\tau \rho \equiv \tanh r$ is a candidate to the proper distance between fundamental particles, just as $e^{t-\tau} \equiv 1/\sqrt{1-\mathcal{R}^2} = \cosh r$ is a plausible relationship of frame time t to proper time τ. Hence, if Bondi & Gold, and Hoyle, want to retain $d\tau^2 - e^{2\tau}d\rho^2$ as a fundamental invariant of their model, in face of the definitions $\rho \equiv \sinh r\, e^{-t} \equiv \tanh r\, e^{-\tau}$, they shall have to accept:

(13)
$$d\tau = dt - dr\, \tanh r \quad \Downarrow \quad e^\tau d\rho = dr - dt\, \tanh r \quad *$$
$$d\mathcal{T}^2 = d\tau^2 - e^{2\tau}d\rho^2 = \{ \tfrac{d\mathrm{T}^2 - d\rho^2}{(1-\mathrm{T})^2} \} = \tfrac{dt^2 - dr^2}{\cosh^2 r}$$

Generalizing these to (1x3) dimensional timespace we find the following three metrics, of which the first one comprises the public flat 3-space of a universe expanding with $\mathcal{S}(\tau) = e^\tau$, whereas the second one contains the public flat 3-space of atoms shrinking with $\mathcal{C}(\mathrm{T}) = 1-\mathrm{T}$, and the third one is closest to represent the private 3-spaces of the standard frames of **SR**:

(14a)
$$d\mathcal{T}^2 = d\tau^2 - e^{2\tau}\{d\rho^2 + \rho^2(d\theta^2 + \sin^2\theta\, d\phi^2)\}$$

(14b)
$$= \tfrac{1}{(1-\mathrm{T})^2}[\,d\mathrm{T}^2 - \{d\rho^2 + \rho^2(d\theta^2 + \sin^2\theta\, d\phi^2)\}]$$

(14c)
$$= [\,dt^2 - \{dr^2 + \sinh^2 r(d\theta^2 + \sin^2\theta\, d\phi^2)\}]\tfrac{1}{\cosh^2 r}$$

$d\mathcal{T}^2 = [\,dt^2 - \{dr^2 + \sinh^2 r(d\theta^2 + \sin^2\theta\, d\phi^2)\}]/\cosh^2 r$ does not compete well with the standard invariant of **SR** which is the much simpler one: $d\mathcal{T}^2 = dt^2 - dr^2 - r^2(d\theta^2 + \sin^2\theta\, d\phi^2)$. This rather serious problem is caused by the external factor $\cosh^{-2} r$ which is much less akin to the **LT** of **SR** than to the **VT** (Voigt transformations) of some competing aether theory.

Therefore, it seems worth while to search for alternative "steady state" models that are not at variance with $d\mathcal{T}^2 = dt^2 - dr^2$. As a step on the way, I shall propose the model below where (15b) & (15c) follow from (15a) by the above definitions $\rho \equiv \sinh r\, e^{-t} \equiv \tanh r\, e^{-\tau}$:

(15a)
$$d\mathcal{T}^2 = dt^2 - \{dr^2 + \sinh^2 r(d\theta^2 + \sin^2\theta\, d\phi^2)\}$$

(15b)
$$= [\,d\tau^2 - e^{2\tau}\{d\rho^2 + \rho^2(d\theta^2 + \sin^2\theta\, d\phi^2)\}]\tfrac{1}{1 - e^{2\tau}\rho^2}$$

(15c)
$$= \tfrac{1}{(1-\mathrm{T})^2}[\,d\mathrm{T}^2 - \{d\rho^2 + \rho^2(d\theta^2 + \sin^2\theta\, d\phi^2)\}]\tfrac{1}{1 - \rho^2/(1-\mathrm{T})^2}$$

This, at the very least, is compatible with the standard invariant $d\mathcal{T}^2 = dt^2 - dr^2$ of **SR**. But $d\tau$ & $d\mathrm{T}$ do not represent cosmic time since an external factor applies to both (16b) & (16c). This clearly shows that (15b) & (15c) do not conform to the **RWM** for expanding space.

*
$$e^\tau d\rho = dr\, \cosh^{-2} r - d\tau\, \tanh r = dr\, \tfrac{(1+\sinh^2 r)}{\cosh^2 r} - dt\, \tanh r = dr - dt\, \tanh r$$

$$=//=$$

Mogens True Wegener

6.   A NEW STEADY STATE MODEL

Let us make a fresh start by adopting $d\mathcal{T}^2 = dt^2 - dr^2$ of **SR**. As we need not accept that standard frames are flat, it seems that we are free to assume that the world is better described, when referring to frame time t, by assuming the 3-space of standard frames to be hyperbolic:

(16) $$d\mathcal{T}^2 \equiv dt^2 - \{dr^2 + sinh^2 r\,(d\theta^2 + sin^2\theta\,d\phi^2)\}$$

Now the shortcomings alluded to in §5 can be remedied by adopting the following definitions:

(17) $$\rho \equiv sinh\,r\,e^{-t} \equiv 2\,tanh\tfrac{r}{2}\,e^{-\tau} \equiv [2]\mathcal{R}e^{-\tau}$$

Interpreting $\mathcal{R}$ as twice the distance to *the midway particle* between two fundamental observers, we shall choose $\mathcal{R} \equiv 2\,tanh\tfrac{r}{2} \to 2$ rather than $\mathcal{R} \equiv tanh\tfrac{r}{2} \to 1$, deleting the factor $[2]$ in (17).

From the above definitions we immediately derive the relationships following below:

(18) $$\mathcal{R} = e^{\tau}\rho \Rightarrow \dot{\mathcal{R}}/\mathcal{R}$$

(19) $$e^t d\rho = dr\,cosh\,r - dt\,sinh r = dr - d\tau\,sinh\,r$$

For fundamental observers, $d\rho = 0$; hence $v \equiv dr/dt = tanh r$, $w \equiv dr/d\tau = sinh r$, whence:

(20) $$cosh\,r = \frac{1}{\sqrt{1-v^2}} = \gamma_r\ .\ sinh r = \frac{v}{\sqrt{1-v^2}} = v\,\gamma_r$$

As shown in the preceding section, the definitions $\rho \equiv sinh\,r\,e^{-t} \equiv tanh\,r\,e^{-\tau}$ might be used to change the old steady state metric (14a), $d\mathcal{T}^2 = d\tau^2 - e^{2\tau}\{d\rho^2 + \rho^2(d\theta^2 + sin^2\theta\,d\phi^2)\}$, into two other metrics (14b) & (14c) where neither τ, nor T, were apt to represent a cosmic time; the third metric was furthermore marred by an external factor that prevented a reduction to **SR**. Fortunately, the alternative definitions $\rho \equiv sinh\,r\,e^{-t} \equiv 2\,tanh\tfrac{r}{2}\,e^{-\tau}$ work much better since, as we have just seen, they allow us to preserve the γ-factor, so characteristic for **SR**.

It is also evident that for $d\theta = d\phi = 0$ (16) reduces to $d\mathcal{T}^2 = dt^2 - dr^2 = dt^2/\gamma^2 = d\tau^2$. Thus we have tested our basic assumption, viz., that the master clocks of fundamental particles always keep the same *Cosmic Rhythm*, $d\mathcal{T} = d\tau = invar.$, the core of a *Cosmic Time*.

Note added 2021:

So far, I have been unable to devise a plausible **RW**-compatible metric for expanding space:

$$e^{t-\tau} = cosh^2\tfrac{r}{2} \Downarrow\ e^{\tau}\rho = 2\,tanh\tfrac{r}{2}$$
$$e^t d\rho = dr\,cosh\,r - dt\,sinh r = dr - d\tau\,sinh\,r$$
$$dr = d\tau\,sinh\,r + e^t d\rho = (d\tau\,2\,tanh\tfrac{r}{2} + e^t d\rho)\,cosh^2\tfrac{r}{2} = \frac{e^{\tau}\rho\,(d\tau + \frac{d\rho}{\rho})}{1 - e^{2\tau}\rho^2}$$
$$dt = d\tau + dr\,tanh\tfrac{r}{2} = d\tau + dr\tfrac{1}{2}e^{\tau}\rho = \frac{d\tau\,(1 - e^{2\tau}\rho^2) + \frac{1}{2}e^{2\tau}\rho^2\,(d\tau + \frac{d\rho}{\rho})}{1 - e^{2\tau}\rho^2} = \frac{d\tau\{1 - \frac{1}{2}e^{2\tau}\rho^2(1 - \frac{d\rho}{\rho d\tau})\}}{1 - e^{2\tau}\rho^2}$$
$$d\mathcal{T}^2 = dt^2 - dr^2 = (dt + dr)(dt - dr) =$$
$$\left\{\frac{d\tau\{1 - \frac{1}{2}e^{2\tau}\rho^2(1 - \frac{d\rho}{\rho d\tau})\}}{1 - e^{2\tau}\rho^2} + \frac{d\tau e^{\tau}\rho\,(1 + \frac{d\rho}{\rho d\tau})}{1 - e^{2\tau}\rho^2}\right\}\left\{\frac{d\tau\{1 - \frac{1}{2}e^{2\tau}\rho^2(1 - \frac{d\rho}{\rho d\tau})\}}{1 - e^{2\tau}\rho^2} - \frac{d\tau e^{\tau}\rho\,(1 + \frac{d\rho}{\rho d\tau})}{1 - e^{2\tau}\rho^2}\right\}$$
$$= \frac{d\tau^2[\{1 - \frac{1}{2}e^{2\tau}\rho^2(1 - \frac{d\rho}{\rho d\tau})\}^2 - e^{2\tau}\rho^2(1 + \frac{d\rho}{\rho d\tau})^2]}{(1 - e^{2\tau}\rho^2)^2} \underset{-d\rho = \rho\,d\tau}{=}\ d\tau^2$$

That $dr = 0$ for $d\rho = -\rho\,d\tau$ means that we follow a series of receding FPs crossing a fixed distance r_o.

Therefore the 3-space of my new world-model is not expanding, but fixed, i.e., stationary.

$$=//=$$

Non-Standard Relativity

7.   CONCLUSION

A very simple alternative construction of my new model, ignoring important details, since the metric is only valid for fundmental particles, might run as follows, writing $c = unity$, using $t \equiv \frac{1}{2}(t_3+t_1)$ & $r \equiv \frac{1}{2}(t_3-t_1)$ for *frame-time* & *frame-distance*, resp., and postulating:

$$dt/d\mathcal{T} = cosh(r/r_o) \ . \ dr/d\mathcal{T} = sinh(r/r_o)$$

From these two simple differential equations all important consequences follow in due order:

$$d\mathcal{T}^2 = dt^2 - dr^2 \ . \ v \equiv dr/dt = tanh(r/r_o)$$
$$dt/d\mathcal{T} = 1/\sqrt{1-v^2} \equiv \gamma \ . \ dr/d\mathcal{T} = v/\sqrt{1-v^2} = v\gamma$$
$$1+z(t) \equiv (dt+dr)/d\mathcal{T} = exp\{r(t)/r_o\} = d\mathcal{T}/(dt-dr)$$
$$1+z(t) = exp(r/r_o) = e \iff r = r_o \equiv t_o \equiv unity$$
$$\mathcal{R} \equiv 2\,tanh(\tfrac{r}{2}) \implies \dot{\mathcal{R}} \equiv d\mathcal{R}/d\mathcal{T} \propto \mathcal{R}$$
$$\mathcal{H} = \dot{\mathcal{R}}(\tau)/\mathcal{R}(\tau) = const.$$

From the Hubble proportionality $\mathcal{H}$, velocity-space being hyperbolic (see Ungar 2008), it is natural to conclude that position-space must be hyperbolic too; so we shall postulate:

$$d\mathcal{T}^2 = dt^2 - dr^2 - sinh^2 r\,(d\theta^2 + sin^2\theta\,d\phi^2)$$

Here, $d\mathcal{T}^{-1}$ may be interpreted as representing an all-embracing **cosmic rhythm**, and we then only have to agree about the arbitrary choice of an universal time *zero,* in order possess a full-fledged **cosmic time** overcoming the reservations expressed by Bondi [1959[2], p.70].

A feature of my new metric is that the element $d\mathcal{T}$ of timespace can be interpreted as the fundamental element of a *cosmic supertime*, as suggested by my mentor André Mercier.

What is also new as compared to the old model is that our metric obeys to the *no-horizon postulate* of Milne; so it survives the number-redshift statistics that refuted the old one.

$$=//=$$

CHAPTER 5

BIG BANG VERSUS STEADY STATE
ARE THESE INCOMPATIBLE IDEAS?

=//=

Revised version (2015,2020,2021) of a paper to the PIRT Conference
'Mathematics, Physics and Philosophy in the Interpretations of Relativity Theory'
Loránd Eötvös University, Budapest 2007.

=//=

Summary

In the present paper it is shown how it
is possible to utilize LT', the differential LT,
as a point of departure for deriving three new
"steady state" models of the universe which are
at variance with the Robertson Walker Metric
but fulfil Milne's cosmological principle.

=//=

Contents

=//=

A.   INTRODUCTION

The γ-factor, which is defined as the quotient between an element of frame time dt and an element of proper time $d\tau$, is the most noticeable consequence of special relativity *(SR)*.

A standard clock passing along a series of slave clocks, distributed over the co-moving ("stationary") frame of an observer, and synchronized in the conventional way to the master clock of that observer, will thus appear retarded according to: $\gamma = dt/d\tau = \sqrt{1-dr^2/dt^2}$.

On the other hand, $d\tau^2 = dt^2 - dr^2 = dt'^2 - dr'^2$ is usually seen as a direct consequence of the differential Lorentz Transformations *(LT′)* in agreement with the relativity principle *(RP)* and the principle of a constant light speed, here termed the "light principle" *(LP)*.

How can $d\tau$ be delayed relative to dt and yet be invariant? In order to solve this problem we must pass on to cosmology. One of the first "big bang" models, probably the only one based on the integral Lorentz Transformations is the uniform expansion model of E.A. Milne [1935]. In a paper on Milne's kinematic relativity theory *(KR)*, his former student A.G. Walker [1937] showed how the model can be restated so as to accomodate a cosmic time, $\mathcal{T}$:

$$t = \tau \cosh\sigma \ . \ r = \tau \sinh\sigma$$
$$d\mathcal{T}^2 = dt^2 - dr^2 = d\tau^2 - \tau^2 d\sigma^2$$
$$d\sigma \to 0 \ . \Rightarrow . \ d\tau \to d\mathcal{T}$$

This result inspired him to devise a method for the development of models incompatible with *LT*, and thus to develop his own version of the famous Robertson-Walker metric *(RWM)*:

$$(1) \qquad\qquad d\mathcal{T}^2 = d\tau^2 - \mathcal{S}^2(\tau)\, d\sigma^2 = invar.$$

Here τ is a universal parameter, $\mathcal{S}(\tau)$ is a universal scale function and σ a fixed ("co-moving") coordinate characterizing one fundamental observer relative to another.

Probably the Milne model * is unique in this respect that it is the only model to combine *LP* in the strict sense with *RWM*. Thus, when Bondi & Gold presented their "steady state" *(SS)* model, they explicitly based it on *RWM*, noticing that their model is not compatible with *LT*.

The scale function $\mathcal{S}$ of their *SS* universe being $\mathcal{S}(\tau) \equiv e^\tau$, it is easy to demonstrate that the model of Bondi & Gold with standard definitions of t & r is incompatible with *LT*:

$$d\mathcal{T}^2 = d\tau^2 - e^{2\tau} d\sigma^2 = (dt^2 - dr^2)(1 - \tanh^2 r)$$

So we may ask if we can devise a new *SS* model by means of *LP & LT′*, the differential *LT*, using the method of Milne, instead of basing it on *RWM*, using the method of Walker.

This confronts us with a choice between $d\mathcal{T}^2 = dt^2 - dr^2$, in combination with *LT′*, and $d\mathcal{T}^2 = d\tau^2 - e^{2\tau} d\sigma^2$, in combination with some other and hitherto unexplored transformations.

Hence our problem: Is a cosmology based on *LT′*, the differential *LT*, at all feasible?

*   **Note added 2021**: The Milne cosmology of *KR* is radically different from the Newtonian model of Milne & McCrea [1934] which, like the similar one of Landsberg & Evans [1979], was never meant to stand up to observation. But a Wikipedia article on *"The Milne Model"* does not even mention *KR*; the Milne model is here confused with a *FLRW* model devoid of matter.

=//=

Mogens True Wegener

1.   A SIMPLE DERIVATION OF LT

A simple way of obtaining the Lorentz Transformations *(LT)* of Special Relativity *(SR)* - not rigorous, but illuminating - would begin with the Galileo Transformations *(GT)*, unprimed coordinates referring to an observer O and primed to another observer O', taking O & O' to be in collinear motion with uniform relative velocity v, as reckoned from O to O':

$$t' = t \ . \ x' = x-vt \ . \ y' = y \ . \ z' = z$$

GT conform to the relativity principle *(RP)*, but not to that of a constant light speed *(LP)*. Thus, if O coincides with O' at the event $(t_o, x_o, 0, 0) = (t'_o, x'_o, 0, 0) = (0, 0, 0, 0)$, a light wave emerging from that event could not be spherical with respect to both observers at the same time, as equation $c^2t^2 = x^2+y^2+z^2$ is not transformed into the similar equation $c^2t'^2 = x'^2+y'^2+z'^2$. So we cannot put $c \equiv c'$; this shows that neither can we synchronize clocks using reflected light, nor can we interpret spatial distance as light-time measured by reflected radar signals.

If, by contrast, we insist on $c \equiv c'$, in *apparent* agreement with the results of observation and experiment, we have to modify *GT* accordingly. However, we may preserve the standard convention $v' = -v$, now enlightened by the fact that the numerical value of both velocities are expressible as the same fraction of the speed of light: $|v|/c = |v'|/c$. A further simplification is obtained if we put $c \equiv c' \equiv 1$, thus interpreting all speeds as fractions of the speed of light. How should *GT* be modified? The only parameter we dispose of is $|v'| = |v| = const.$

So let us define a new function $\gamma \equiv \gamma(v)$. According to *RP*, γ would have to be invariant. The simplest assumption is that γ only affects transformations in the direction of the x-axis, leaving the other spatial dimensions unaffected. But t being involved in the transformation of x to x', it may affect t' too, thus necessitating a renunciation of the classical transformation $t = t'$. An additive constant would hardly do the job. Let us try if γ could be a multiplicative factor:

(2) $$x' = \gamma(x-vt) \ . \ y' = y \ . \ z' = z \ . \ x = \gamma(x'-v't')$$

Assuming $v' = -v$, it must be done this way, for v is the velocity of a fix-point in the frame of O' as calculated by O: $dx' = \gamma(dx-vdt) = 0 \Rightarrow dx = vdt$, just as v' is the velocity of a fix-point in the frame of O as calculated by O': $dx = \gamma(dx'-v'dt') = 0 \Rightarrow dx' = v'dt'$. From $x' = \gamma(x-vt)$ combined with $x = \gamma(x'+vt')$ we then derive the transformation for time:

$$x = \gamma(x'+vt') = \gamma\{\gamma(x-vt)+vt'\} \Rightarrow t' = \gamma\{t-x(1-\gamma^{-2})/v\}$$

In case of light propagation, as seen above: $t^2-(x^2+y^2+z^2) = t'^2-(x'^2+y'^2+z'^2) = 0$. Further, all velocities being fractions of the velocity of light, the following must hold in general:

$$t^2-(x^2+y^2+z^2) = t'^2-(x'^2+y'^2+z'^2)$$

On account of $y' = y$ & $z' = z$ we accordingly obtain: $t^2-x^2 = t'^2 = x'^2$. It is therefore evident that the temporal co-ordinates t & t' must transform in the same way as the spatial ones x & x':

(3) $$t' = \gamma\{t-x(1-\gamma^{-2})/v\} = \gamma(t-vx)$$

From this it is easy to derive a precise expression for the γ-factor; we obtain:

(4) $$\gamma = 1/\sqrt{1-v^2}$$

Hence we may say that the eqs. $x' = \gamma(x-vt)$ & $x = \gamma(x'+vt')$ contain the very germ of *LT*.

Non-Standard Relativity

2.   THE IMPORTANCE OF THE γ-FACTOR

Granted **LP** and $c \equiv c' \equiv 1$, the natural definition of coordinates is the standard one:

$$\tau_3 \equiv t + x \quad \Downarrow \quad \tau_1 \equiv t - x$$
$$t = \tfrac{1}{2}(\tau_3 + \tau_1) \; . \; x = \tfrac{1}{2}(\tau_3 - \tau_1)$$
$$\tau_4' \equiv t' + x' \quad \Downarrow \quad \tau_2' \equiv t' - x'$$
$$t' = \tfrac{1}{2}(\tau_4' + \tau_2') \; . \; x' = \tfrac{1}{2}(\tau_4' - \tau_2')$$

Why 'τ'? The point is that τ_i & τ_i' are read directly off the *master-clocks* of O & O', resp., denoting the proper times of those clocks, whereas t & t' are read off the *slave-clocks* fixed to the co-moving frames F_o & $F_{o'}$ of O & O', resp., denoting the standard frame times of F_o & $F_{o'}$ resp. t & t'. So *master-clocks* show *proper times* τ while *slave clocks* show *frame times* t.

Imagine a zig-zag signal passing to & fro between O & O' directly, without any delay:

$$... < \tau_1 < \tau_2' < \tau_3 < \tau_4' < ...$$

These are epochs taken to be read off the master-clocks of O & O' at the events of reflection. Granted **RP**, τ_4' must be the same function of τ_3 as τ_3 of τ_2', and as τ_2' of τ_1, viz., $\tau_{i+1} \equiv \psi(\tau_i)$:

(5) $$\tau_4' = \psi(\tau_3) \; . \; \tau_3 = \psi(\tau_2') \; . \; \tau_2' = \psi(\tau_1)$$

The signal-function ψ forms the core of the Milne-Whitrow derivation of **LT**, hailed as *transcendental* by J.R. Lucas [1971]. Taking the relative velocities of O & O' to be uniform and reciprocal, $v \equiv -v'$, the function ψ must be linear, of the form: $s\tau + k$. The Doppler Shift (DS), which we identify with its derivative: $d\psi/d\tau = s$, must then be both constant and reciprocal:

(6) $$1 + z \equiv d\tau_3/d\tau_2' = d\tau_2'/d\tau_1 \equiv 1 + z'$$

(7) $$1 + z = \sqrt{d\tau_3/d\tau_1} = \sqrt{\tfrac{dt + dx}{dt - dx}} = \sqrt{\tfrac{1 + v}{1 - v}} = e^{arth\, v}$$

Synchronizing the master-clocks of O & O' to read $\tau = \tau' = 0$ by coincidence, we choose $k \equiv 0$ for signals exchanged directly between O & O'; hence $\tau_3/\tau_2' = s = \tau_2'/\tau_1$. Further:

$$\tau_2' = \sqrt{\tau_3 \tau_1} = \sqrt{(t + x)(t - x)} = t\sqrt{1 - x^2/t^2}$$
$$\tau_3 = \sqrt{\tau_4' \tau_2'} = \sqrt{(t' + x')(t' - x')} = t'\sqrt{1 - x'^2/t'^2}$$

Thus, in case of "photon" transmission: $\frac{x}{t} = \frac{x'}{t'} = 1$, the element of proper time $d\tau$ will be zero even though the element of frame time dt is not. Further, for the relative motion of O & O':

$$x' = 0 \Rightarrow : x = vt \Rightarrow dt = d\tau'/\sqrt{1 - v^2} = \gamma\, d\tau'$$
$$x = 0 \Rightarrow : x' = v't' \Rightarrow dt' = d\tau/\sqrt{1 - v'^2} = \gamma\, d\tau$$

This proves that the master-clock of O' showing $d\tau'$ will appear to be retarded relative to a series of slave-clocks distributed on the x-axis of O, just as the master-clock of O showing $d\tau$ will appear to be retarded relative to a series of slave-clocks distributed along the x'-axis of O'.

So proper time differs from frame time, there is no surprise in this, cf. H. Arzeliés [1966]. What *is* surprising, however, is that in case of energy change, due to the performance of work, the retardation will be absolute. Thus a moving clock will lack behind a resting clock of the same construction if it returns to its point of departure after having completed a circuit in space.

Nevertheless, as far as inertial collinear motion is concerned, we can reduce **LT** to **GT** for any pair of observers. This will be demonstrated in the following section.

Mogens True Wegener

3.   THE FORMAL REDUCTION OF LT TO GT

With $\gamma = 1/\sqrt{1-v^2}$, as known, the standard **LT** can be stated in the following form:

(8) $\qquad\qquad x' = \gamma\,(x - vt) \ . \ y' = y \ . \ z' = z \ . \ x = \gamma\,(x' - v't')$

Insert the values of t & t' from $\tau \equiv t - \frac{x}{v}(1-\gamma^{-1}) \equiv t' - \frac{x'}{v'}(1-\gamma^{-1})$ into **LT**, then, lo and behold: we immediately obtain something which is surprisingly similar to the **GT** of classical physics!

(9) $\qquad\qquad x' = x - v\,\tau\,\gamma \ . \ y' = y \ . \ z' = z \ . \ x = x' - v'\tau\,\gamma$

But is τ a time displayed on a physical clock? Take the derivative of τ, and see what we get:

$$d\tau \equiv dt - \frac{dx}{v}(1-\gamma^{-1}) \equiv dt' - \frac{dx'}{v'}(1-\gamma^{-1})$$

(10) $\qquad d\tau \underset{dx\,=\,v\,dt}{=} dt/\gamma = \sqrt{dt^2 - dx^2} = \sqrt{dt'^2 - dx'^2} = dt'/\gamma \underset{dx'\,=\,v'dt'}{=} d\tau'$

So the clock sought for has the same rate as the master-clocks of F & F'. Further, $x' = x$ for $\tau = 0$; this proves that the new clock will agree with the two master-clocks by coincidence. These are the criteria of *clock-congruence*, or *synchrony*. Thus τ is the common time of F & F'. That such a time exists is the great *no!-no!* of **SR**. But "c'est ne pas tout", to quote Poincaré. In fact, as shown by J. Winnie [1970], a simple additive adjustment of time-zero for each single slave clock will suffice to ensure that all the slave-clocks distributed over the entire co-moving frames of F & F' will agree by coincidence - and will continue to do so for ever after!

From the point of view of an observer M situated precisely midway between F & F', i.e., $MF \equiv MF'$ - and such *a midway observer* and his co-moving frame *can always be constructed* - it is evident both that the master-clocks of F & F' are in perfect synchrony, $d\tau = d\tau'$, and that their co-moving slave-clocks, after adjustment of their time-zeros, keep exactly the same rate. Hence, *to make clocks in uniform collinear motion tick in unison is not a question of changing their clock mechanisms, but only a question of adjusting their time-zeros properly.* Apparently the trick cannot be performed with more than one pair of frames at a time. But look at this:

Consider the case of three or more observers $F, F', F'', \ldots$ in uniform collinear motion. If they coincide at the same event, $t = t' = t''$, we can always devise an adjustment of time-zeros for their slave-clocks in order to make them all agree, without changing their mechanisms. In fact, we need no more than a single slave-clock in each fix-point of their co-moving frames if only each of these clocks serves as time-keeping mechanism for any number of pointers, each of these pointers being adjusted with its own time-zero, depending on the relative velocity of that observer with respect to whose co-moving slave-clocks it is intended to agree.

Now consider instead a triply infinite (∞^3) set of observer-particles, or particle-observers. Assume that the structure of the whole set is defined by the following property: for any non-collinear triple of particle-observers belonging to the set there is a fourth one, member of the set, which is the mid-way particle of the first three, so that it remains *equidistant* from those three. My point, then, is that such a set is a substratum of fundamental observers in the sense of Milne, thereby fulfilling his specific formulation of the *cosmological principle (CP)*; cf. North [1965]. This being the case, all proper distances $\mathcal{R}$ between the members of such a substratum will be subject to the same scale function $\mathcal{S}(\tau)$, with the same universal time τ as argument.

According to Milne & Whitrow [1949], the accept of a unique substratum of fundamental particles / observers / monads is the only way to avoid the socalled "clock paradox" of **SR**.

Non-Standard Relativity

4.   FROM "BIG BANG" TO "STEADY STATE"

As we have already seen, Walker rendered the basic invariant of Milne's **KR** thus:

$$dT^2 \;=\; dt^2 - dr^2 \;=\; d\tau^2 - \tau^2 d\sigma^2$$

This he generalized to the standard metric of modern cosmology **(RWM)**, cf. North [1965]:

$$dT^2 \;=\; d\tau^2 - \mathcal{S}^2(\tau)\, d\sigma^2 \;=\; invar.$$

Here τ is a universal parameter, argument in the scale factor $\mathcal{S}(\tau)$ and, for $d\sigma = 0$, identical to the invariant cosmic time T which is the common proper time read off the master-clocks of all so-called fundamental observers, members of the substratum. For $d\sigma = 0$ we obtain:

(11)
$$\sigma = \int_{\tau_1}^{\tau_2} \frac{d\tau}{\mathcal{S}(\tau)} = \int_{\tau_2}^{\tau_3} \frac{d\tau}{\mathcal{S}(\tau)} = const.$$

Walker saw σ as a fixed "co-moving" coordinate characterizing one fundamental observer relative to another fundamental observer, their proper distance being defined as $\mathcal{R}(\tau) \equiv \mathcal{S}(\tau)\sigma$. The curvature of 3-space, which is latent in the element $d\sigma$, he left undetermined to begin with. A time scale which eliminates the spatial expansion by the factor $\mathcal{S}(\tau)$, so that all distances between the fundamental observers remain constant, is definable by $T = \int d\tau / \mathcal{S}(\tau) + const.$ When we know the form of the scale-factor $\mathcal{S}(\tau)$, we can deduce the cosmological red-shift, $1 + z(\tau) \equiv \mathcal{R}(\tau_3)/\mathcal{R}(\tau) = \mathcal{S}(\tau_3)/\mathcal{S}(\tau)$, from $\sigma = const.$ in eq.(11) for any standard **RWM**.

The short terms, "big bang" and "steady state", were both coined by Hoyle. The first real "big bang" model was devised by a catholic priest, Lemaître, who spoke of a "primeval atom". Milne preferred to speak of *a transcendent point-event*. According to the Friedmann-Lemaître equations of General Relativity **(GR)**, the **BB** model of Milne and the **SS** model of Bondi & Gold are both devoid of matter. The tacit presupposition, of course, is the general validity of the field equations of **GR**. But neither Milne, nor Bondi & Gold, accepted those equations. *

In fact, **KR** was meant by Milne to replace both **SR** & **GR**. With **KR**, Milne devised his own very ingenious theory of gravitation. As I have argued in an earlier paper, Wegener [2000], his idea can be described as "turning Mach's principle upside down": instead of trying in vain to explain inertia by reducing it to a kind of gravitation, he proposed to explain gravitation as a kind of inertia; this he did by reducing it to local deviations from global (cosmic) symmetry. From this point of view, a cosmological model is primarily described by its scale function, so there is no question of a global gravitational field acting as a "brake" on universal expansion, hence also not of taking the universe to be filled up with dark anti-gravitational energy.

The analogy between $d\tau^2 = dt^2 \gamma^{-2} = dt^2 - dr^2$ and $dT^2 = d\tau^2 - \mathcal{S}^2(\tau)d\sigma^2$ has been subject to conjecture by G.J. Whitrow [1961], another student of Milne's, who suggested that **RWM** may be derivable from the strong **LP** of **SR**. However, his procedure is not convincing. Moreover, it turns out that some interesting metrics, viz., that of the "steady state" universe of Bondi & Gold, and those of two other kindred models, are in conflict with his assumption.

These models - as defined by the scale factors $\mathcal{S}_1(\tau) \equiv e^\tau$, $\mathcal{S}_2(\tau) \equiv sh\tau$, $\mathcal{S}_3(\tau) \equiv ch\tau$ - present us with the choice between preserving the standard **RWM** form and discarding **LP** as a principle of universal validity, or preserving the universal validity of **LP** and discarding **RWM**. Instead of following Bondi & Gold by choosing the first option, I shall prefer the second one.

*   **NB: confer the note on p.64, added 2021.**

5.   THE COSMOLOGICAL PRINCIPLE OF MILNE

Let us state a few definitions, unprimed entities referring to O and primed ones to O':

$$v^2 \equiv v_x^2 + v_y^2 + v_z^2 \ . \ v'^2 \equiv v_x'^2 + v_y'^2 + v_z'^2$$
$$v \equiv dr/dt, \ v_x \equiv dx/dt \ . \ v_y \equiv dz/dt, \ v_z \equiv dz/dt$$
$$v' \equiv dr'/dt' , \ v_x' \equiv dx'/dt' \ . \ v_y' \equiv dz'/dt' , \ v_z' \equiv dz'/dt'$$

Making use of **LT'**, Milne calculated the velocity distribution of particles in a kinematic substratum as it is displayed to two fundamental observers, O & O', "at rest" in the substratum. Since O & O' both observe the same set of objects, viz., the substratum, they must agree about:

(12) $$f_o(v_x, v_y, v_z)\, dv_x dv_y dv_z = f_{o'}(v_x', v_y', v_z')\, dv_x' dv_y' dv_z'$$

The **CP**, which is taken to hold for all fundamental observers, members of the substratum, but neither for arbitrary objects, nor for accidental observers not belonging to the substratum, can be viewed as a strong universal **RP** supporting the definability of an invariant cosmic time. Milne himself interpreted **CP** as a principle stating the formal identity of the functions f_o & $f_{o'}$:

(13) $$f_o \equiv f_{o'} \equiv f$$

In order to investigate the consequences of the above identity, he made use of **LT'**:

$$dx' = \frac{dx - v_{oo'}dt}{\sqrt{1 - v_{oo'}^2}} \ . \ dy' = dy \ . \ dz' = dz \ . \ dt' = \frac{dt - v_{oo'}dx}{\sqrt{1 - v_{oo'}^2}}$$
$$v_x' = \frac{v_x - v_{oo'}}{1 - v_x v_{oo'}} \ . \ v_y' = \frac{v_y(1 - v_{oo'}^2)}{1 - v_y v_{oo'}} \ . \ v_z' = \frac{v_z(1 - v_{oo'}^2)}{1 - v_z v_{oo'}}$$

Applying partial differentiation to **LT'** he derived the following provisional results [1948,§52]:

$$f(v_x, v_y, v_z) = f(v_x', v_y', v_z') \frac{\partial v_x' \partial v_y' \partial v_z'}{\partial v_x \partial v_y \partial v_z}$$
$$\frac{\partial v_x' \partial v_y' \partial v_z'}{\partial v_x \partial v_y \partial v_z} = \frac{(1 - v_{oo'}^2)^2}{(1 - v_x v_{oo'})^4}$$
$$f(v_x, v_y, v_z) = f\left(\frac{v_x - v_{oo'}}{1 - v_x v_{oo'}}, \frac{v_y(1 - v_{oo'}^2)}{1 - v_y v_{oo'}}, \frac{v_z(1 - v_{oo'}^2)}{1 - v_z v_{oo'}} \right) \frac{(1 - v_{oo'}^2)^2}{(1 - v_x v_{oo'})^4}$$

The most general solution of these functional equations Milne [1935] showed to be:

$$f(v_x, v_y, v_z)\, dv_x dv_y dv_z = \mathcal{B}\, \gamma^4\, dv_x dv_y dv_z$$
$$\gamma \equiv \frac{1}{\sqrt{1 - v_x^2 - v_y^2 - v_z^2}} \ . \ \mathcal{B} = const.$$

Expressed in polar coordinates, with $d\omega$ denoting a small solid angle, his result can be written:

(14) $$f(v, \omega)\, v^2 dv\, d\omega = \mathcal{B}\, \gamma^4 v^2 dv\, d\omega$$

Now, passing from this velocity distribution to the corresponding positional distribution, Milne applied the basic property of his uniformly expanding model, viz., the constancy of the relative velocities between all fundamental observers (members of the subtratum) pairwise.

But at this point we must deviate from his procedure, the relative velocities between pairs of fundamental observers in our model being no longer constant, but increasing with distance.

So, in order to proceed, we shall exploit the property $v \equiv dr/dt = tanh\, r$ implied by:

$$d\mathcal{T} = dt/cosh\, r = dr/sinh\, r = invar.$$

Non-Standard Relativity

6.   A NEW MODEL OF CONTINUED CREATION *(CC)*

The *basic equations* at the end of §6 are easily integrated; we just present the result:

(15) $$\underline{\mathcal{R}(\mathcal{T}) \equiv e^{\mathcal{T}}\rho = e^{t}\rho / ch^{2}\frac{r(t)}{2} = 2\,tanh\frac{r(t)}{2}} \quad \cdot \quad \left(\begin{array}{l} \mathcal{T}=invar.\\ \rho=const. \end{array} \right)$$

The *cosmic redshift* of our model is the usual Doppler Shift (*DS*) of *SR* (cf. ch.3, §6):

$$1+z(t) = e^{r(t)} = \frac{1+tanh\frac{1}{2}r(t)}{1-tanh\frac{1}{2}r(t)} = \frac{1+\frac{1}{2}\mathcal{R}(\mathcal{T})}{1-\frac{1}{2}\mathcal{R}(\mathcal{T})}$$

A *natural calibration of units*, $r/r_{o} = 1 \neq \mathcal{R}/\mathcal{R}_{o} = 1$, is made in the following way:

$$r = r_{o} \equiv 1 \Leftrightarrow \mathcal{R} = 2\,tanh\frac{1}{2} \Leftrightarrow t-\mathcal{T} = ln\,cosh^{2}\frac{1}{2} \Leftrightarrow z(t) = e-1$$

$$\mathcal{R} = \mathcal{R}_{o} \equiv 1 \Leftrightarrow r = 2\,arth\frac{1}{2} \Leftrightarrow t-\mathcal{T} = ln\,2 \Leftrightarrow z(\mathcal{T}) = \frac{1+\frac{1}{2}}{1-\frac{1}{2}}-1 = 2$$

So far our model fulfils the **dimensional postulate** of Milne [1948, §72]: *no dimensional constant of nature should be allowed to enter the definition of the kinematic substratum.*

Considering a series of fundamental particles we have $d\rho \neq 0$ and, by differentiation:

$$e^{\mathcal{T}}d\rho = d\mathcal{R} - \mathcal{R}\,d\mathcal{T} \ . \ \ d\mathcal{T} = dt - tanh\frac{r}{2}\,dr$$

$$dr' \equiv e^{t}d\rho = cosh\,r\,dr - sinh\,r\,dt = dr - sinh\,r\,d\mathcal{T}$$

But, for $d\rho = 0$, we recover the characteristics of a true model of continued creation, cf. eq.(16):

(16) $$\underline{d\mathcal{T} = d\mathcal{R}/\mathcal{R} = dr/sinh\,r = dt/cosh\,r = dt/\gamma = invar.}$$

(17) $$\underline{\underline{\mathcal{R}(\mathcal{T}) \equiv e^{\mathcal{T}}\rho \Rightarrow \mathcal{H} \equiv \dot{\mathcal{R}}/\mathcal{R} = 1}}$$

My *CC*-model shows some structural similarity to the *BB*-model of Milne, but the crucial difference is that my *CC*-world is stationary whereas his *BB*-world is expanding; furthermore, his 3-space is flat, whereas I assume the 3-space of the substratum to be hyperbolic.

Milne made a very illuminating distinction between the universe as *world-map* and the universe as *world-view* (literally, he spoke of 'world-picture', but my notion is slightly different). Defined as simultaneous co-existence, the universe just now should be conceived as world-map. Perceived as momentary appearance, the universe shows itself to the observer as world-view.

The World-Map of my new CC-universe describes the-universe-as-it-is-in-itself as an unobservable unity of simultaneous co-existence in an infinite hyperbolic timespace:

(18) $$\underline{d\mathcal{T}^{2} = dt^{2} - [c_{o}^{-2}]\,ds^{2} = invariant} \ . \ \underline{ds^{2} = dr^{2} + sinh\,r^{2}(d\theta^{2} + sin^{2}\theta d\phi^{2})}$$

The World-View of my new CC-universe depicts the-universe-as-it-appears-to-us as a visible show of shells of varying age making a finite pseudo-sphere of $\mathcal{R} \equiv 2$ in flat 3-space, disclosing a contraction of objects with distance from the origo (cf.p.5, Escher-picture!):

(19) $$\underline{[c_{o}^{2}]dt^{2} \doteqdot [c_{o}^{2}]\,d\mathcal{T}^{2} + ds^{2}} \ . \ \underline{ds^{2} \doteqdot \{d\mathcal{R}^{2} + \mathcal{R}^{2}(d\theta^{2} + sin^{2}\theta\,d\phi^{2})\}/(1-\frac{\mathcal{R}^{2}}{4})^{2}}$$

$$=//=$$

7.   TWO ASYMPTOTIC APPROXIMATIONS

The Model M_2 approximates our continuous creation model M_1 from a "Fierce Blow". Model M_2 is constructed in the following way: Adopting LT' - the differential Lorentz Group as entailed by the strong LP - we assume these time dependent velocity-distance relations:

(20) $$\rho \equiv \sinh r / \sinh t \equiv 2 \tanh \tfrac{r}{2} / \sinh \tau \equiv \mathcal{R} / \sinh \tau$$

Our all-embracing cosmic time $\mathcal{T}$ can now be written:

(21) $$\tau = arsinh\{\sinh t / \cosh^2 \tfrac{r}{2}\} = invar.$$
$$\rho = \mathcal{R}(\tau)/\sinh \tau = \mathcal{R}(\tau_3)/\sinh \tau_3 = const.$$
$$1 + z(\tau) = \sinh \tau_3 / \sinh \tau \underset{\tau \to \infty}{\longrightarrow} e^{r(\tau)} = \frac{1 + \tfrac{1}{2}\mathcal{R}(\tau)}{1 - \tfrac{1}{2}\mathcal{R}(\tau)}$$

The corollaries of (20) yields some interesting relations:
$$\sinh t\, d\rho = \cosh r\, dr - \sinh r\, dt \coth t = dr - \sinh r\, d\tau \coth \tau$$
$$v \equiv dr/dt \underset{d\rho=0}{=} \tanh r / \tanh t = \tanh r \sqrt{1 + \sinh^2 t}/\sinh t = \sqrt{\sinh^2 r + \rho^2}/\cosh r$$
$$\gamma_v \equiv 1/\sqrt{(1 - v^2)} \underset{d\rho=0}{=} 1/\sqrt{1 - (\sinh^2 r + \rho^2)/\cosh^2 r} = \cosh r / \sqrt{1 - \rho^2}$$
$$\mathcal{H}_2(\mathcal{T}) \equiv \dot{\mathcal{R}}_2(\mathcal{T})/\mathcal{R}_2(\mathcal{T}) \propto \coth \mathcal{T} \underset{\tau \to \infty}{\longrightarrow} \mathcal{H}_1(\mathcal{T})$$

We see that, near time zero, the velocities of dispersion far transcend that of light thereby obviating the supposed need for inflation in a very natural way.

The Model M_3 approximates our model M_1 of continuous creation by a "Gentle Flow". Model M_3 is constructed in a similar manner: Adopting LT' - the differential Lorentz Group as entailed by the strong LP - we assume these time dependent velocity-distance relations:

(22) $$\rho \equiv \sinh r / \cosh t \equiv 2 \tanh \tfrac{r}{2} / \cosh \tau \equiv \mathcal{R} / \cosh \tau$$

Our all-embracing cosmic time $\mathcal{T}$ can now be written:

(23) $$\tau = arcosh\{\cosh t / \cosh^2 \tfrac{r}{2}\} = invar.$$
$$\rho = \mathcal{R}(\tau)/\cosh \tau = \mathcal{R}(\tau_3)/\cosh \tau_3 = const.$$
$$1 + z(\tau) = \cosh \tau_3 / \cosh \tau \underset{\tau \to \infty}{\longrightarrow} e^{r(\tau)} = \frac{1 + \tfrac{1}{2}\mathcal{R}(\tau)}{1 - \tfrac{1}{2}\mathcal{R}(\tau)}$$

The corollaries of (22) also yields some interesting relations:
$$\cosh t\, d\rho = \cosh r\, dr - \sinh r\, dt \tanh t = dr - \sinh r \tanh \tau\, d\tau$$
$$v \equiv dr/dt \underset{d\rho=0}{=} \tanh r \tanh t = \tanh r \sqrt{\cosh^2 t - 1}/\cosh t = \sqrt{\sinh^2 r - \rho^2}/\cosh r$$
$$\gamma_v \equiv 1/\sqrt{(1 - v^2)} \underset{d\rho=0}{=} 1/\sqrt{1 - (\sinh^2 r - \rho^2)/\cosh^2 r} = \cosh r / \sqrt{1 + \rho^2}$$
$$\mathcal{H}_3(\mathcal{T}) \equiv \dot{\mathcal{R}}_3(\mathcal{T})/\mathcal{R}_3(\mathcal{T}) \propto \tanh \mathcal{T} \underset{\tau \to \infty}{\longrightarrow} \mathcal{H}_1(\mathcal{T})$$

We see that, at time zero, the entire universe is completely stationary, forming what has sometimes been described as a "cosmic egg".

$$=//=$$

8.   CONSIDERATIONS OF ENERGY

The kinematic substratum functions as a *compass of inertia* (Weyl, Gödel) by defining the states of rest and motion in the universe. While fundamental observers may be considered to be locally at rest, all other particles not members of the substratum - let us call them accidental - are distinguished by their motion. Now the substratum is dense, cf. the midway property of §3. If an accidental particle A of restmass m_o is passing a fundamental observer F with velocity $\vec{v}_{FA} \equiv \vec{v}_{FF'}$ at instant t_F, we have by the same token found that other fundamental observer F' relative to which it is instantaneously at rest, their distance being $\vec{r}_{AF'} \equiv \vec{r}_{FF'}$ at the instant t_F.

Hence the instantaneous state of motion of an accidental particle is fully specified by two fundamental observers: that with which it coincides, and that with respect to which it is at rest. Now, according to CP, all fundamental particles are equivalent. Granted that the energy in any volume of fixed size, is constant, cf. the energy principle (PCE), and assuming the classical equivalence of velocity of escape from a gravitational potential, $\frac{1}{2}v_\infty^2 = -\varphi = Gm_o/r$, what appears as the kinetic energy of A relative to F, $T_{FA} = m_o\gamma_{\vec{v}} = m_o/\sqrt{1-\vec{v}^2}$, must likewise appear as the dynamic, or potential, energy of A relative to F', $-V_{F'A} = m_o\gamma_{\vec{\varphi}} = m_o/\sqrt{1+2\vec{\varphi}}$. So we shall take T+V to be a basic constant for fundamental observers in accordance with:

(24)
$$T = m_o[c^2](\gamma_{\vec{v}}-1) = m_{\vec{v}}-m_o$$

(25)
$$-V = m_o[c^2](\gamma_{\vec{\varphi}}-1) = m_{\vec{\varphi}}-m_o$$

$$dT = F\,dr = (\tfrac{dp}{dt})dr = \tfrac{d}{dt}(m_o\gamma_v v)dr$$
$$= m_o\{v\tfrac{d\gamma_v}{dt}+\gamma_v\tfrac{dv}{dt}\}dr = m_o\{v^2 d\gamma_v+\gamma_v v\,dv\}$$
$$= m_o\{v^2 d\gamma_v+\gamma_v^{-2} d\gamma_v\} = m_o[c^2]d\gamma_v = dm_{\vec{v}}$$
$$-dV = F\,dr = \{\tfrac{d}{dr}(m_o\gamma_\varphi)\}dr = m_o[c^2]d\gamma_\varphi = dm_{\vec{\varphi}}$$

(26)
$$H \equiv T+V = m_{\vec{v}}-m_{\vec{\varphi}} = m_o(\gamma_{\vec{v}}-\gamma_{\vec{\varphi}}) = const.$$

(27)
$$L \equiv T-V = m_{\vec{v}}+m_{\vec{\varphi}}-2m_o = m_o(\gamma_{\vec{v}}+\gamma_{\vec{\varphi}}-2)$$

Further, assuming the principle of least action (**PLA**), and using the above Lagrangian, we get *a variational principle describing the observed perihelion displacement of Mercury*:

(28)
$$\delta\int_{t_1}^{t_2} L\,dt = \delta\int_{t_1}^{t_2}(m_{\vec{v}}+m_{\vec{\varphi}})dt = 0$$
$$\Rightarrow \tfrac{d}{dt}\big(\tfrac{\partial m_{\vec{v}}}{\partial \dot{q}_i}\big) - \tfrac{\partial m_{\vec{\varphi}}}{\partial q_i} = 0$$

Following Prokhovnik [1988], a unit-rod moving in a substratum ("ether") is reduced by:

(29)
$$c_{v,\theta}^{-1} = \tfrac{1}{2}(c_{\rightarrow}^{-1}+c_{\leftarrow}^{-1}) = \frac{\sqrt{1-v^2\sin^2\theta}}{1-v^2}$$

Thus, due to the local asymmetry introduced by the motion, the longitudinal "speed" of a photon will be $c_{v,0} = 1-v^2$, its transversal "speed" being $c_{v,\frac{\pi}{2}} = \sqrt{1-v^2}$. By analogy we then assume:

(30)
$$c_{v,\theta}^{-1} = \frac{\sqrt{1-v^2\sin^2\theta}}{1-v^2} \simeq c_{\varphi,\theta}^{-1} = \frac{\sqrt{1+2\varphi\sin^2\theta}}{1+2\varphi}$$

Now $\delta\int_{t_1}^{t_2}\frac{2r}{c_\varphi}dt$ for $\theta \simeq 0$ yields *the observed delay of light-signals* reflected from a planet while *the observed deflection of light rays* near a massive body is found by a Fermat principle:

(31)
$$\delta\int_{t_1}^{t_2}\frac{dr}{c_\varphi} = \delta\int_{r_1}^{r_2}\frac{\sqrt{1+2\varphi\sin^2\theta}}{1+2\varphi}dr = 0$$

So *the γ-factor* (§§ 2-3) plays a crucial rôle in our derivation of observational effects.

Mogens True Wegener

O.   CONCLUSION

For astronomical purposes, *the concept of light-time distance r is particularly convenient.* In this measure, which relates to frame-time t, the Hubble function takes on the non-linear form:

$$v/r \underset{t \to \infty}{\to} tanh\, r/r$$

This is a very definite prediction. Expressed in proper distance $\mathcal{R}$ and proper time $\mathcal{T}$ it becomes:

$$d\mathcal{R}/\mathcal{R}\, d\mathcal{T} \underset{\mathcal{T} \to \infty}{\to} unity$$

Our notion of proper distance, $\mathcal{R} \equiv 2\, th\frac{r}{2} \to e^{\mathcal{T}}\rho$, may look strange, due to the factor 2. But it has a simple explanation, being *the distance between two fundamental observers, F & F', as defined in the frame of their midway-particle, M.* Put $v_{FM} = tanh\, r_{FM} = tanh\frac{1}{2}r_{FF'} \underset{r \to \infty}{\to} 1$ and $v_{MF'} = tanh\, r_{MF'} = tanh\frac{1}{2}r_{FF'} \underset{r \to \infty}{\to} 1$; then, just as $v_{FM}+v_{MF'} \underset{r \to \infty}{\to} 2$, we have:

$$\mathcal{R}_{FF'} \equiv 2\, tanh\tfrac{1}{2}r_{FF'} = tanh\, r_{FM}+tanh\, r_{MF'} \to 2$$

Thus, while light-time distance r is simply additive, proper distance $\mathcal{R}$ adds up like $2\, th\frac{r}{2}$. So our three models MM_{1-3} display a density increasing with distance according to world-view. In this way our models conform to the *dictum* of cardinal Nicholas of Cusa [ca.1450]:

The world machinery is as if it had its center everywhere and circumference nowhere, its circumference and center being no other than God who is everywhere and nowhere.

From every possible point of view, defined by reference to a single fundamental observer, the universe is interpretable as a steady stream, confined within the limits of a sphere, which may be stationary (our "*Steady State*" model M_1), exploding from a transcendent singularity (our "*Fierce Blow*" model M_2) or emanating from a static state (our "*Gentle Flow*" model M_3).

All models MM_{1-3} are non-Friedmannian and conform to Milne's no-horizon principle. So they avoid the usual problems associated with the standard *RW* models (flatness, horizons). There is no reason to invoke the idea of inflation. Further, being comprised within the confines of a pseudo-sphere with radius $\mathcal{R}_o = 2$, a constant universal mass and energy is definable.

With our world M_1, we are confronted with the picture of an eternal universe where the energy lost at the periphery is compensated by a steady gain in energy at the center, from which a steady aethereal stream of matter and light is pouring out towards all possible directions.

MM_{2-3} are just different approximations to the same overall view of continuous creation. For this reason our new *CC* models MM_{1-3} can be claimed to represent the true synthesis of the opposite cosmological views of two ancient philosophers: *Parmenides & Herakleitos.*

$$=//=$$

CHAPTER 6

NEW AXIOMS FOR COSMOLOGY
FROM SPACETIME VIA TIMESPACE TO SUPERTIME

*Revised version (2015,2021) of a paper presented at the PIRT Conference
'Mathematics, Physics and Philosophy in the Interpretations of Relativity Theory'
Loránd Eötvös University, Budapest 2009.*

=//=

Summary:

*In the present paper it is shown how it is possible
by means of a time-based concept of equidistance to
construct a spatial geometry for relativistic cosmology.
In analogy to a sphere defined as the geometrical site for
all those points which are equidistant from a given point,
we construct a plane as the site for all those points that
are equidistant from two points, and a line as the site
for all those points equidistant from three points.*

*Having defined parallellity and perpendicularity
we proceed to define the cosmic substrate in a way
analogous to the cosmological principle of the Cusan;
assuming that we can always construe the center point
for any three non-collinear members of this substrate,
we can prove simultaneity to be universally transitive
for all members of the substrate, if the simultaneity
is defined indirectly bys means of equidistance.*

=//=

Mogens True Wegener

PREAMBLE

An axiomatization of relativitic cosmology may be construed with various aims in mind. One goal is to codify pet ideas and entrenched theories thus giving boost to scientific dogma. Quite another is it to invent a model depicting some basic traits of the universe in order to test that structure against experience. It is the second purpose that has motivated this paper.

Instead of anticipating what is written in my paper, I prefer to say a little more about why I find it important to put focus on such a very general structure displaying such particular traits. Like the philosopher Bergson, I always felt that the so-called "spatialization of time" is a fatal misconception on behalf of science, indeed the ultimate disaster. The idea of a "block universe" wherein nothing happens, everything existing of eternity though in a timeless way, is foreign to experience, and the fantasy of "time-warps" and "spacetime tunnels" leading to other worlds is, in my opinion, nothing but idle speculation bolstered up with subtle mathematics.

Eddington once said: "In physics everything depends on the insight with which the ideas are handled *before* they reach the mathematical stage". These words are true, wise and pertinent; but, like Eddington, most physicists still turn to Einstein as if *his* ideas were the highest wisdom. Now, according to Einstein, time is what is read off our clocks and, if our clocks are retarded, time itself must be dilated (this inference was made by Minkowski, then adopted by Einstein). Further, Maxwell's equations invoke a constant c that is naturally interpreted as a limiting speed, viz., that of light. What is more natural, then, than to insist that time is another form of space?

Long forgotten is the doctrine of Descartes, that mind and matter are two different kinds of "being" with one property in common, viz., their temporal duration or "durée", cf. Bergson. Equally eclipsed by oblivion is the insight of Kant that, whereas the intuition of *space* yields the *geometric* form of all *external* experience, the intuition of *time* yields the *arithmetic* form not only of internal experience, but of all experience, *external as well as internal*. It was precisely this insight which inspired Hamilton to invent his quaternions, so fruitful to physics.

But is it not "a reactionary move" to propose an axiomatics to revitalize those outmoded concepts of absolute simultaneity and universal time so characteristic of classical physics? Well, I have not the slightest doubt that professor Nemeti and his followers will consider it that way. As far as I understand, they want to base their axiomatics on the socalled light-cone geometry. The point to be noticed here is that the opening angle of the cone to its axis is representative for the velocity of light: for the one-way velocity of light, that is.

Now the standard version of the special theory of relativity *(SR)* is distinguishable from other competing theories not only by its two basic principles, that of relativity and that of the constancy of the velocity of light, but similarly by the fact that its definition of simultaneity at a distance is determined purely by convention. This, of course, entails that its definition of the one-way velocity of light must be conventional too! That this is so was shown by John Winnie [1970], and in another way by my friend Peter Øhrstrøm [2000].

SR is obviously equivalent to a host of theories, each with its own one-way light-speed. So, according to SR, whereas the average or two-way velocity of light is a universal constant, the one-way velocity of light - hence the opening angle of its light cones - is wholly arbitrary. This, as far as I see it, puts the geometric light-cone enterprise completely in jeopardy.

Non-Standard Relativity

To this argument it might be objected that, recalling Reichenbach's ϵ-constant, Einstein's choice of $\epsilon = \frac{1}{2}$ is unique in the sense that it accords with the value obtained by infinitely slow clock transport. It may be adduced that Malament [1997] has presented a famous argument to the effect that the value $\epsilon = \frac{1}{2}$ is the only one that allows for a reversal of the direction of time. Finally, this value appears to be indispensable to the standard definition of a reference frame.

But these objections are nothing but subterfuges. Just like the definition of simultaneity at a distance, the one-way velocity of light for a certain distance is either conventional, or it is not. If we stick to *SR* it is certainly conventional, that is, indifferent to experiment and observation, and then the standard concept of a reference frame is nothing but a convenient fiction, or a fake. If it can be shown that it is not conventional, we shall have to search for an alternative theory, involving a different concept of reference frame than the usual one of standard relativity.

It is commonplace that physicists - instead of taking incompatibilities and contradictions for what they are, namely, incompatibilities and contradictions - try by all means to evade them by making distinctions so subtle that no one can sort them out. This applies in particular to the attempts at unifying *GR* with *QM*. Very few seem willing to face the fact that *GR* and *QM* are incompatible and that all attempts to unify them inevitably lead us into a mess of contradictions.

Now dr. Rowlands, in his conference paper, suggests that gravity may be instantaneous. This, in my opinion, is a suggestion that might well be true and should not be lightly dismissed. He admits that "there is, therefore, an incompatibility between the spacetime structure of our observations (supposing it to be that of standard *SR*) and the spacetime structure we require to set up our gravitational equations" but, adds he, "this is not an uncommon occurrence in physics and there is a well known solution". To a logician such words sounds a little distressing.

I agree with dr. Rowlands that "quantum gravity is a meaningless idea". But I believe the same holds of his own proposal that, if we apply a Lorentzian spacetime to a system subject to non-local gravitation, this incompatibility can be circumvented by introducing something called "fictitious effects" to compensate for an otherwise outright contradiction. It is clear that in order to vindicate the view that gravitational effects are instantaneous, the notion of simultaneity at a distance must be uniquely definable, and this it is not with standard *SR*. Here logic holds sway, and I see no reason to endorse current attempts to modify basic logic to comply with physics. But dr. Rowland's *distinction between instantaneous and observational structure* is certainly relevant, cf. my own distinction between *World-Map* and *World-View*, this book ch.5 §6.

Another of my friends, professor Selleri, points to what I think may be a better way out. Having in his forthcoming new book on *Weak Relativity* assembled and discussed a number of observations and experiments in support of the revolutionary idea that simultaneity, after all, is absolute, he proceeds to develop some new inertial transformations to accomodate that idea. Accepting the evidence here accumulated, I agree that we shall have to search for a new theory. An essential characteristic of the inertial transformations proposed and generalized in his book is that the dependence of the temporal coordinate on the longitudinal spatial one so distinctive of standard *SR* is here suspended, so that only the usual γ-factor of *SR* remains relevant.

The same remark applies to another non-standard theory developed by my friend Tom Phipps jr., who in his monumental book *Heretical Verities* [1986] has construed a full-fledged socalled neo-Hertzian electrodynamics in due respect to almost all current relativistic evidence. Regrettably, however, these two heretical spirits do not agree concerning the spatial coordinate.

Furthermore, they are both reluctant to accept the full validity of the strong relativity principle. In my own opinion this is a very serious drawback of their respective non-standard theories and if I shared their view, I might not have been that eager to invite our guest professor Ungar.

In any case I agree with professor Ungar that the standard Lorentz transformations should be applicable to all equivalent observers in a universe which is in a state of uniform expansion. The latter condition, of course, imposes a problematic restriction on our cosmological theories; so we may have to search for other transformations if the expansion, as it seems, is accelerating. but the need for new transformations is also urgent if the observers involved are not equivalent. Now the statement that the expansion is accelerating must refer to an accepted standard, and the usual standard is given by the size of our own body, i.e., by the size of its constituting atoms.

This standard being extrapolated *ad infinitum* in space we get our usual reference frames, and if such frames are synchronized by the usual convention, our choice of an origo seems free. What I surmise is that this is a mistake involving the dissolution of classical simultaneity; so if, with Selleri and Phipps, we want to keep simultaneity absolute, this mistake should be avoided. I agree that all this may sound strange. How can I both accept a weak concept of relativity for the reason that it allows simultaneity to be absolute and yet endorse a strong relativity principle? An explanation is obviously urgent. But my point is that a distinction must be made.

A way out of the impasse was shown long ago by E.A. Milne, in his *Kinematic Relativity*. In this classical monograph he proposed a world-model whose structure is determined by an infinite substratum of perfectly equivalent fundamental particles exposed to uniform dispersion. This substratum, the members of which are subject to the *strong relativity* principle, was then supposed to be covered by a layer of accidental particles subject to a weaker kind of relativity. Milne's universe thus consists of two sets of particles: the substratum of *fundamental particles* subject to *strong relativity* as formulated in the principle of cosmic isotropy, and a further layer of *accidental particles* breaking the cosmic symmetry, hence subject to *weak relativity*.

Milne denounced the Friedmann equations of **GR**, thus his model is not Friedmannian. Instead of assuming gravitation, hiding it in the curvature of space, it was his aim to deduce it. He began by deriving the Lorentz transformations and showed them to hold good between the comoving frames of all fundamental observers, *implying that each frame has its natural origo in the only fundamental observer at rest in that frame*, all others receding with uniform velocity. His assistant Walker then showed the standard coordinates of **SR** to be transmutable into some other coordinates mapping the universe to be in a state of uniform expansion in cosmic time.

As I see it, the difference between the many *private* Lorentzian 3-spaces, each frame correlated with its own Einsteinian t-time, and the single, unique and *public* Robertson-Walker 3-space correlated with the cosmic T-time of Walker, is also the key to an understanding of the ingenious non-commutative non-associative algebra invented and studied by professor Ungar.

The Milne model, which is very clean, was later adopted by Törnebohm and Prokhovnik; it can be described as having its origin not in a "big bang", but in a transcendent singularity. Applying his cosmological principle to the Lorentz transformations, and making an important distinction between *world map* (the universe as it is in itself at an instant of cosmic time) and *world view* (the universe as it appears to an observer, sliced up in different temporal layers), cf. the famous one of Kant between "das Ding an sich" and "das Ding für uns", Milne was able to describe precisely how his world model of uniform dispersion relates to observation.

How did this remarkable feat fall into oblivion, so that Milne and his kinematic relativity theory is hardly mentioned in more recent expositions of cosmology? Well, one explanation is that the fame of Einstein had reached mythological dimensions overshadowing that of all others. In our modern culture, what is greater than to become depicted on the side of a shopping bag? Every child knows the name of Einstein, but who cares about that of Plato, except philosophers? Another explanation is that Milne's gravitational theory did not stand up to observational test, but was surpassed and excelled in that respect by Einstein's. But this is just another myth.

For the sake of fairness one must allow a theory to become developed and, as reported by Whittaker in his *History of the Theories of Aether and Electricity vol.2*, Walker early suggested a variational principle based on *SR* that almost reproduced the observational resuslts of *GR*; so he issued a much ignored warning: don't accept a theory with only scant experiental support. It later became almost a sport to construe formulae based on *SR* and imitating the results of *GR*. I have given two myself, and the issue has been studied in depth by Rowlands [1994, 2007].

I only want to urge the ingenuity of Milne's basic idea: to derive gravity from asymmetry. We have recently learned much about broken symmetries leading to differentiation of forces: strong, weak, electromagnetic, etc., but hitherto noone has succeeded to incorporate gravitation. The project of "quantum gravity" is still unsolved, despite the endless efforts of mathematicians. But a solution may be close at hand and much simpler than expected: just follow Milne!

According to Milne's kinematic relativity theory, the substratum is an ideal, infinite and dense set of equivalent particles subject to hydrodynamic continuity and obeying Hubble's law. Since he assumed the universe to be in a state of *uniform dispersion* - expressed with the metric of Walker: in a state of *uniform expansion* - his universe embodies a perfect *compass of inertia*, to use a felicitous phrase of Weyl. This means that it is impossible to construe an inertial frame which is not in the end identifiable with the frame comoving with some fundamental observer.

But, as already said, *only one fundamental observer can be at rest in an inertial frame: this observer constituting the natural origo of that frame*, so we are not free to choose another. The substratum, as told, is covered by a layer of accidental particles. Now, in contrast to the fundamental particles which are at rest, each one in its own comoving frame, the position and velocity of an accidental particle is completely described by reference to two fundamental ones: that with which it is momentarily coincides and that relative to which it is momentarily at rest, and the closer these two are to each other, the more the particle tends to fundamental status.

Milne's model, like all other models of the universe, is subject to conservation of energy. Now an accidental particle, due to its motion, possesses a certain amount of kinetic energy as estimated by that fundamental observer with which it momentarily coincides. But fundamental observers are equivalent. The accidental particle therefore possesses exactly the same amount of energy as estimated by any other fundamental observer, and only its form may seem different. How, in particular, must its energy be perceived by that observer relative to which it is at rest? Only one possibility is left open to him: the energy must appear as being potential, or dynamic. *Thus, what to the first observer appears as **inertia**, to the second one must appear as **gravity**!* With Milne, *cosmic time* is conditioned by *symmetry*, *gravity* is caused by *asymmetry*.

I agree with professor Ungar that there is an intimate connexion between special relativity and hyperbolic geometry. The same does dr. Barrett, and so did Milne. But there is a difference. Whereas professor Ungar ascribes a hyperbolic character to the *velocity space* of *SR*, dr. Barrett also consider its *position space* to be hyperbolic. At this point, I side with dr Barrett.

Mogens True Wegener

1.   INTRODUCTION.

The Greeks prescribed that proofs of geometry should be made solely by means of a ruler and a pair of compasses. In this paper it will be shown how the ruler can be dispensed with, and how all concepts of geometry can be constructed by means of "a temporal pair of compasses", in a way reminiscent of, but also different from, the method devised by Georg Mohr [1672].

Geometry treats of spatial structure and, according to Descartes, space is real, because extension is a sign of substance on a par with cogitation. But Leibniz denied this, arguing that extension, being infinitely divisible, cannot be substantial. According to Leibniz, space is not real, but neither is it illusive; rather it is *well-founded appearance*. To this view he was inspired by Plato who, in his *Timaios*, held that space is neither pure concept (*idéa*), nor pure appearance (*fainómenon*), but "something in between" so that, seeing it "as in a dream", we cannot tell what it is, but feel that it has to be, in order for events to take place. Thus, to Plato, space is necessary in order to give room for events, whereas time alone is "an image in motion of eternity".

Contrary to common prejudice the Platonic view of space is supported by modern science Thus, in their excellent monograph *On General Relativity*, Mercier, Treder & Yourgrau [1979] unanimously held that "*there is no such thing as real space*" (p.134). Further Mercier [2000], in a more philosophical article, advocated the view that 'spacetime' should be reconstructed as 'timespace', or 'supertime'. But, more than half a century before this, Milne already proposed the same idea, first in several papers, partly written in collaboration with his student Whitrow, then in his own ingenious *Kinematic Relativity* [1948/1951], wherein he constructed an entirely new cosmology from first terms based on time as its fundamental concept. This cosmology was described by Merleau-Ponty [1965] as "*a Leibnizian monadology*" translated into mathematics. The present paper may be read as a modest attempt to elaborate on the very same idea.

According to Whitrow [1972], "Leibniz's principle (of a pre-established harmony, MTW) is equivalent to the postulate of a single universal time; we must therefore discard this principle - i.e., if we are to reconcile Leibniz's way of regarding time with Einstein's theory of relativity". But why attempt to reconcile Leibniz with Einstein? As shown by Walker - another student of Milne's who, independently of Robertson, invented the relativistic standard metric of universes that are everywhere isotropic - such universes not only *allow*, but outright *demand*, a universal time parameter determining the spatial scale factor and, as demonstrated by Törnebohm [1963], even special relativity gives place to two distinct concepts of time and two different definitions of simultaneity: one relative, another absolute. Therefore, as argued by myself, Wegener [2004], we have every reason to search for a new formalization of special relativity consistent with the idea of a Cosmic Time, even if this goes against "the spirit of Einstein".

=//=

2.   FORMAL PRESENTATION

In what follows I have found it more important to convey an impression of the main ideas rather than to offer an impeccable presentation of some full-fledged formalization.

The axiomatics is sketched by means of 1st order predicate logic (***FOL***) or quantification theory (***QT***) supplemented with basic set theory, $\forall$ being the universal quantor, $\exists$ the existential quantor, $\Rightarrow$, $\wedge$, $\vee$, $\neg$, $\Leftrightarrow$ being the constants of implication, conjunction, disjunction, negation, and equivalence, and $=$, and $\simeq$, being the relations of identity and similarity. $\forall x : Ax \Rightarrow Bx$, $\exists x : Ax \wedge Bx$ are well-formed formulae. $\{., x, y, z, .\}$ and $\{x \,|\, (description)\}$ are simple sets, $\in$ denoting membership. Pauses are marked by points. Quantors are omitted from definitions. Definitions are indicated by $\equiv$ which, depending on context, means either $=_{df}$ (i.e. identity: "*is/are*") or $\Leftrightarrow_{df}$ (i.e. equivalence: "*iff*"). Finally, $\rightarrow$ denotes a signal, or the exchange of a single "photon" (a one-one relation between an emission at one particle and an absorption at another particle), while $\preceq$ betokens the causal relation of precedence or succession of signals.

Df.1: *The Universe.*
$$\boldsymbol{Univ} . \equiv . \left\{ \mathrm{P}^i \,|\, \mathrm{P}^i \equiv \{.., P_1^i, P_2^i, P_3^i, ..\} \right\}$$
Translation: The universe is the set of all existing *points*, or *particles* - or *particle-observers*, or *observer-particles*, or **monads** - which themselves are (or consist of) sets of observed events.

AX.1 *Causation by Signals.*
$$\forall \mathrm{P}, \forall \mathrm{P}', \forall P_a : P_a \in \mathrm{P}. \Rightarrow . \exists 1 P_e' : P_e' \in \mathrm{P}' \wedge P_e' \rightarrow P_a$$
Translation: For any two particles P & P', and for any event P_a in P (absorption of a "photon"), there was one, and only one, event P_e' in P' (emission of a "photon") which was the cause of P_a. *Comment:* One must distinguish *backwards inference*, i.e. the **explanation** of a *present effect* by its *past cause*, from *forwards inference*, i.e. the **prediction** of a *future effect* by its *present cause*. In order to avoid discussing probabilities we put focus on explanation rather than prediction. Ax.1 says that all events, e.g., state P_a of P, were caused by preceding states of other particles. So there was *no first event - any possible first event was beyond all possible experience*.

Df.2 *Successive Causation.*
$$P_e \preceq P_a' . \equiv . \exists \mathrm{P}, \exists \mathrm{P}' : P_e \in \mathrm{P} \wedge P_a' \in \mathrm{P}' \wedge P_e \rightarrow ... \rightarrow P_a'$$
Translation: For any pair of events, P_e in P and P_a' in P', we shall say that the event of absorption P_a' *was* caused (not: *will be* caused) by the event of emission P_e, iff a signal was emitted by P at event P_e and, possibly after transmission by other events, observed by P' at the event P_a'.

AX.2 *Transitivity of Succession.*
$$P_1' \preceq P_2'' \wedge P_2'' \preceq P_3''' . \Rightarrow . P_1' \preceq P_3'''$$
Translation: If P_1' caused P_2'' and if P_2'' caused P_3''', then P_1' caused P_3'''.

Df.3 *Momentary Coincidence.*
$$\mathrm{PP}'(P_r') = 0 . \equiv . \exists P_e, \exists P_a : P_e, P_a \in \mathrm{P} \wedge P_e \preceq P_r' \preceq P_a \wedge P_e = P_a$$
Translation: The ("light-time") distance between P and P' at event P_r' in P' was zero iff there were events P_e and P_a in P, so that P_e caused P_r' which caused P_a, but P_e coincided with P_a.

Mogens True Wegener

Df.4 *Permanent Coincidence.*
$$P = P' \; . \; \equiv \; . \; \forall P'_r : PP'(P'_r) = 0$$
Translation: Two particles, P & P', are *identical* iff their momentary distance is always zero.

Df.5 *Momentary Equidistance.*
$$PP'(P'_r) = PP''(P''_r) \; . \; \equiv \; . \; \exists P_e, \exists P_a : P_e, P_a \in P \wedge P_e \to P'_r \to P_a \wedge P_e \to P''_r \to P_a$$
Translation: The light-time distance from P to P' at the event P'_r in P' equalled the light-time distance from P to P'' at the event P''_r in P'' iff both events of reflection were caused by the same event of emission P_e in P and coincidentially caused the same event of absorption P_a in P.
*Comment: This temporal definition of **equidistance** is basic to our axiomatization of geometry;* being comparable to the span of "a pair of compasses", it justifies our use of that metaphor. The notion of equidistance presupposes that "photons" are transmitted with invariant "two-way" speed in all directions, i.e., *isotropically*. We do *not* assume a constant "one way" light speed.

Df.6 *Permanent Equidistance.*
$$PP' = PP'' \; . \; \equiv \; . \; \forall P'_r, \forall P''_r : PP'(P'_r) = PP''(P''_r)$$
Translation: The distances from P to P' & P" remain equal iff they are equal at all events.

Df.7 *Momentarily Smaller/Greater Distance.*
$$PP'(P'_r) < PP''(P''_r) \; . \; \equiv \; .$$
$$\exists P_e, \exists P_a, \exists P_{a'} : P_e, P_a, P_{a'} \in P \wedge P_e \to P'_r \to P_a \wedge P_e \to P''_r \to P_{a'}$$
$$\wedge \; \exists P''', \exists P'''_r : PP'''(P'''_r) \neq 0 \wedge P_a \to P'''_r \to P_{a'}$$
Translation: Distance PP' at P'_r in P' was *smaller* than distance PP" at P''_r in P" - equivalently, $PP''(P''_r)$ was *greater* than $PP'(P'_r)$ - iff both reflection events P'_r & P''_r were caused by the same event of emission P_e in P, but caused successive events of absorption P_a & $P_{a'}$ in P, so that a signal emitted at P_a, when reflected at P'''_r by P''' at distance $PP'''(P'''_r) \neq 0$, arrived at $P_{a'}$.

Df.8 *Permamently Smaller / Greater Distance.*
$$PP' < PP'' \; . \; \equiv \; . \; \forall P'_r, P''_r : PP'(P'_r) < PP''(P''_r)$$
Translation: PP' remains smaller than PP" iff it is smaller at all events of reflection.

Df.9 *Optical Lines.*
$$\mathbf{opt}(P'_e \preceq P_a) \; . \; \equiv \; . \; P'_e \to ... \to P_a \wedge P'_e \to P_a$$
Translation: The causal succession $P'_e \preceq P_a$ formed an optical line iff two signals beginning at the same emission event $P'_e \in P'$: the indirect one $P'_e \preceq P_a$, and the direct one $P'_e \to P_a$, both terminated at the same absorbtion event $P_a \in P$; i.e., the optical line was observed at $P_a \in P$.

Df.10 *Light-Cones.*
$$\mathbf{cone}(P_a) \; . \; \equiv \; . \; \{ P^i_e | \mathbf{opt}(P^i_e \preceq P_a) \}$$
Translation: The light-cone observed at $P_a \in P$ was the set of all optical lines terminating at P_a.
Comment: The only way of determing the top angle of past cones is by arbitrary convention; this corresponds to the fact that, in *SR*, "the one-way light-speed" depends on convention.

Df.11 *Observational Perspectives.*
$$\mathbf{persp}(P) \; . \; \equiv \; . \; \{ P_a | \mathbf{cone}(P_a) \}$$

Translation: The observational perspective of P was the set of all light cones terminating in P.
Comment: An observational perspective frames the **world-view** of an observer, cf. ch5 §6.

Df.12 *Spheres.*
$$\boldsymbol{S}(P, P') \equiv \{P^i \mid PP^i = PP' \neq 0\}$$
Translation: A sphere (a spherical surface) is the set (the geometric site) of all those points that keep the same distance to a given fix-point P as does another given point P', different from P.
Comment: If we allowed $PP' = 0$, the sphere $\boldsymbol{S}(P, P')$ would degenerate into a point.

Th.1 *Non-identity of Spheres.*
$$\boldsymbol{S}(P, P') \neq \boldsymbol{S}(P', P)$$
Translation: A sphere about P with radius PP' differs from a sphere about P' with radius P'P.
Proof: $\quad \forall P, \forall P' : P \neq P' \Rightarrow . \boldsymbol{S}(P, P') = \{P^i \mid PP^i = PP'\} \neq \{P^i \mid P'P^i = P'P\} = \boldsymbol{S}(P', P)$

Df.13 *Similarity of Spheres.*
$$\boldsymbol{S}(P, P') \simeq \boldsymbol{S}(P'', P'') . \equiv . \forall P^i, \forall P^j : PP^i = PP' = P''P''' = P''P^j$$
Translation: Spheres are similar if their defining points are equidistant, i.e., their radii equal.

Cor.1 *Similarity of Spheres is Reciprocal.*
$$\boldsymbol{S}(P, P') \simeq \boldsymbol{S}(P', P)$$
Translation: The similarity of different spheres with equal radii is reciprocal.
Proof: Follows immediately from the preceding definition of similarity.

Cor.2 *Similarity of Spheres is Transitive.*
$$\boldsymbol{S}(P, P') \simeq \boldsymbol{S}(P', P'') \wedge \boldsymbol{S}(P', P'') \simeq \boldsymbol{S}(P'', P''') . \Rightarrow . \boldsymbol{S}(P, P') \simeq \boldsymbol{S}(P'', P''')$$
Translation: The similarity of different spheres with equal radii is transitive.
Proof: Follows immediately from the preceding definition of similarity.

Df.14 *Circles.*
$$\boldsymbol{C}(P, P') \equiv \{\boldsymbol{S}(P, P') \cap \boldsymbol{S}(P', P)\}$$
Translation: A circle is definable as the intersection between two similar spheres.
Comment: So, if the radius of $\boldsymbol{C}(P, P')$ were measurable, it would be $|PP'| / \sqrt{2}$.

Th.2 *Circles as Sets.*
$$\boldsymbol{C}(P, P') = \{P^i \mid PP^i = P'P^i = PP'\}$$
Translation: A circle is the set of all points P^i keeping the same distance PP' to both P and P'.
Proof: $\quad \forall P, \forall P', \forall P^i : P \neq P' \neq P^i \Rightarrow \{\boldsymbol{S}(P, P') \cap \boldsymbol{S}(P', P)\} = \{\{P^i \mid PP^i = PP'\} \cap$
$\{P^i \mid P'P^i = P'P\}\} = \{P^i \mid PP^i = PP' \wedge P'P^i = P'P\} = \{P^i \mid PP^i = P'P^i = PP'\}$

Df.15 *Planes.*
$$\boldsymbol{Pl}(P, P') \equiv \{P^i \mid PP^i = P'P^i \neq 0\}$$
Translation: A plane is the set of all points that are pairwise equidistant to two points, $P \neq P'$.
Comment: The defining points do not belong to the plane defined, i.e., $P, P' \notin \boldsymbol{Pl}(P, P')$

Th.3 *Identity of Planes.*
$$\boldsymbol{Pl}(P, P') = \boldsymbol{Pl}(P', P)$$

Translation: The plane between P and P′ is identical to the plane between P′ and P.

Proof: $\quad \forall P, \forall P' : P \neq P' \Rightarrow . \{P^i \mid PP^i = P'P^i\} = \{P^i \mid P'P^i = PP^i\}$

Th.4 *Circles as Subsets of Planes.*
$$\boldsymbol{C}(P, P') \subset \boldsymbol{Pl}(P, P')$$
Translation: The circle defined by P, P′ is a subset of the plane defined by P, P′.

Proof: $\quad \{P^i \mid P \neq P' . \wedge . PP^i = P'P^i = PP'\} \subset$
$$\{P^i \mid P \neq P' . \wedge . PP^i = P'P^i = PP' \vee PP^i = P'P^i \neq PP'\}$$

Th.5 *Circles as Intersections of Spheres and Planes.*
$$\boldsymbol{C}(P, P') . = . \boldsymbol{S}(P, P') \cap \boldsymbol{Pl}(P, P')$$
Translation: The sphere $\boldsymbol{S}(P, P')$ and the plane $\boldsymbol{Pl}(P, P')$ intersect in the circle $\boldsymbol{C}(P, P')$.

Proof: $\forall P, \forall P' : P \neq P' \Rightarrow . \{\{P^i \mid PP^i = PP'\} \cap \{P^i \mid PP^i = P'P^i\}\} \equiv \{P^i \mid PP^i = P'P^i = PP'\}$

Df.16 *Straight Lines.*
$$\boldsymbol{Lin}(P, P', P'') \equiv \{P^i \mid PP' < PP'P'' \wedge P'P'' < P'PP'' \wedge P''P < P'P''P \wedge PP^i = P'P^i = P''P^i\}$$
Translation: A line is the set of all points equidistant from three given points forming a triangle.

Comment: The defining points do not belong to the line defined, i.e.: $P, P', P'' \notin \boldsymbol{Lin}(P, P', P'')$. The definition yields a measure for the curvature (deviation from straightness) of optical lines.

Th.6 *Identity of Lines.*
$$\boldsymbol{Lin}(P, P', P'') = \boldsymbol{Lin}(P', P'', P) = \boldsymbol{Lin}(P'', P, P') =$$
$$= \boldsymbol{Lin}(P'', P', P) = \boldsymbol{Lin}(P', P, P'') = \boldsymbol{Lin}(P, P'', P')$$
Translation: The line defined by P, P′, P″ is identical to that defined by P′, P″, P which is identical to the one defined by P″, P, P′, i.e., a line is indifferent to the order of its defining points.

Proof: $\{P^i \mid PP^i = P'P^i = P''P^i\} = \{P^i \mid P'P^i = P''P^i = PP^i\} = \{P^i \mid P''P^i = PP^i = P'P^i\} =$
$$= \{P^i \mid P''P^i = P'P^i = PP^i\} = \{P^i \mid P'P^i = PP^i = P''P^i\} = \{P^i \mid PP^i = P''P^i = P'P^i\}$$

Th.7 *Lines as Intersecting Planes.*
$$\boldsymbol{Lin}(P, P', P'') = \{\boldsymbol{Pl}(P, P') \cap \boldsymbol{Pl}(P, P'')\}$$
Translation: The line defined by equidistance to P, P′, P″ is identical to the intersection of the plane defined by equidistance to P, P′ and the plane defined by equidistance to P, P″.

Proof: $\forall P, P' : P \neq P' \Rightarrow . \{P^i \mid PP^i = P'P^i = P''P^i\} = \{\{P^i \mid PP^i = P'P^i\} \cap \{P^i \mid PP^i = P''P^i\}\}$

Th.8 *Lines as Subsets of Planes.*
$$\boldsymbol{Lin}(P, P', P'') \subset \boldsymbol{Pl}(P, P')$$
Translation: The line defined by P, P′, P″ is a subset of the plane defined by P, P′.

Proof: $\quad \{P^i \mid PP^i = P'P^i = P''P^i\} \subset \{P^i \mid PP^i = P'P^i = P''P^i \vee PP^i = P'P^i \neq P''P^i\}$

Df.17 *Parallelity of Planes.*
$$\boldsymbol{Pl}(P, P') \parallel \boldsymbol{Pl}(P'', P''') . \equiv . \exists \boldsymbol{Lin}(P^1, P^2, P^3) : P, P', P'', P''' \in \boldsymbol{Lin}(P^1, P^2, P^3)$$
Translation: Two planes are said to be parallel iff their defining points belong to the same line.

Df.18 *Parallelity of Lines.*
$$\boldsymbol{Lin}(P, P', P'') \parallel \boldsymbol{Lin}(P^1, P^2, P^3) . \equiv . \exists \boldsymbol{Pl}(P^a, P^b) : P, P', P'', P^1, P^2, P^3 \in \boldsymbol{Pl}(P^a, P^b)$$
Translation: Two lines are said to be parallel iff their defining points belong to the same plane.

Non-Standard Relativity

Df.19 *Perpendicularity.*
$$\boldsymbol{Pl}(P^1, P^2) \perp \boldsymbol{Lin}(P, P', P'') . \equiv . P^1, P^2 \in \boldsymbol{Lin}(P, P', P'') \vee P, P', P'' \in \boldsymbol{Pl}(P^1, P^2)$$
Translation: The plane defined by P^1, P^2 is said to be perpendicular to the line defined by P, P', P'' (not collinear) iff either: a) the points P^1, P^2 belong to the line defined by P, P', P'', or: b) the points P, P', P'' belong to the plane defined by P^1, P^2; the two appear to be equivalent, but *Comment:* So far, I have been unable to devise an elegangt proof for the equivalence of a) & b).

Df.20 *The Substrate.*
$$\boldsymbol{Subst} \equiv \{\, P \mid \forall P', P'', P'_r, P''_r : PP'(P'_r) = PP''(P''_r) \Rightarrow PP' = PP'' \}$$
Translation: The Substrate is a set of points / particles characterized by the following property: if a member P of the Substrate is ever equidistant to two other members, P' & P'', it always is.

Df.21 *Fundamental Particles* (FP).
$$P = P_F . \equiv . P \in \boldsymbol{Subst}$$
Translation: The particle P is called fundamental, P_F, iff it belongs to the Substrate.
Comment: All FP are said to be *at rest* in the Substrate.

Df.22 *Accidental Particles* (AP).
$$Q = Q_A . \equiv . Q \notin \boldsymbol{Subst}$$
Translation: The particle Q is called accidental, Q_A, iff it does not belong to the Substrate.
Comment: All AP are said to be *in motion* in the Substrate.

AX.3 *The Substrate is not Empty.*
$$\exists P, P', P'', P''' : P, P', P'', P''' \in \boldsymbol{Subst} . \wedge .$$
$$\forall P^1, P^2 : P^1, P^2 \in \boldsymbol{Subst} . \Rightarrow . P, P', P'' \in \boldsymbol{Pl}(P^1, P^2) \Rightarrow P''' \notin \boldsymbol{Pl}(P^1, P^2)$$
Translation: The Substrate contains four FP not belonging to the same plane.

Df.23 *Midway Particles.*
$$P = \boldsymbol{m}(P^1, P^2) . \equiv . PP^1 = PP^2 \wedge P^1 P^2 = P^1 PP^2$$
Translation: P is the midway particle of P^1 and P^2 iff PP^1 equals PP^2 and $P^1 P^2$ equals $P^1 PP^2$.

AX.4 *The Substrate is Dense.*
$$\forall P^1, P^2 : P^1, P^2 \in \boldsymbol{Subst} . \Rightarrow \exists P : P = \boldsymbol{m}(P^1, P^2) \in \boldsymbol{Subst}$$
Translation: Any two FPs have a midway particle which is also an FP.

AX.5 *The Substrate is Infinite.*
$$\forall P^1, P^2 : P^1, P^2 \in \boldsymbol{Subst} . \Rightarrow \exists P^3 : P^3 \in \boldsymbol{Subst} \wedge P^2 = \boldsymbol{m}(P^1, P^3)$$
Translation: For any two FPs there is a 3rd FP so that the 2nd is midway between 1st & 3rd.

AX.6 *Coincidence entails Collapse.*
$$\forall P, P', P'', P''_r : P, P', P'' \in \boldsymbol{Subst} \Rightarrow . PP''(P''_r) = 0 \Rightarrow P'P''(P''_r) = 0$$
Translation: It holds for any three members of the Substrate that if two of them ever coincide at an instant they all coincide at that instant; thus the entire Substrate collapses into a *singularity*.

Df.24 *Isotropy of Perspectives.*
$$\boldsymbol{iso\text{-}persp}\,(P) . \equiv . \boldsymbol{persp}\,(P) . \wedge . \forall P' : \boldsymbol{persp}\,(P) \cap \boldsymbol{Pl}(P, P') . = . C(P, P')\}$$

Translation: The world-view **persp** (P) of an observer P is *isotropic* iff, for all other particles P′, the intersection of **persp** (P) and **Pl**(P, P′) is the circle **C**(P, P′) with center in **m**(P, P′).
Comment: If the **iso-persp** (P) of P is cut by a plane defined by P and some other particle P′ indicating an arbitrary direction in space, the intersection is always circular.

Th.9 *Isotropic Perspectives are Sets of Concentric Spheres.*
$$\textbf{iso-persp}\,(\mathrm{P})\,.\, =\,. \{\mathrm{P}^i \mid \mathrm{P}^i \in \textbf{persp}\,(\mathrm{P}) \cap \textbf{Pl}(\mathrm{P}, \mathrm{P}')\,. \Rightarrow\, \mathrm{P}^i \in \textbf{S}(\mathrm{P}, \mathrm{P}')\}$$
Translation: The isotropic perspective of an observer P is the set of all spheres centered in P.
Proof: By df.24, $\textbf{persp}\,(\mathrm{P}) \cap \textbf{Pl}(\mathrm{P}, \mathrm{P}')\,. =\,. \textbf{C}(\mathrm{P}, \mathrm{P}')$, and by df.14, $\textbf{C}(\mathrm{P}, \mathrm{P}') \subset \textbf{S}(\mathrm{P}, \mathrm{P}')$

AX.7 *The Substrate is Isotropic.*
$$\forall \mathrm{P} : \mathrm{P} \in \textbf{Subst}\,. \Rightarrow\,. \textbf{persp}\,(\mathrm{P}) = \textbf{iso-persp}\,(\mathrm{P})$$
Translation: The observational perspective of a fundamental particle P is always isotropic.
Comment: The Substrate fulfils the *Principle of Cosmic Isotropy*, cf. Cusanus [ca.1450].

Th.10 *Isotropy of Light-Cones.*
$$\forall\, \mathrm{P}, \mathrm{P}', \mathrm{P}^\iota : \mathrm{P}, \mathrm{P}', \mathrm{P}^\iota \in \textbf{Subst} : : \Rightarrow : : \forall P_e, P^\iota_r, P_a : P_e \preceq P^\iota_r \preceq P_a\,. \wedge\,.$$
$$P^\iota_r \in \textbf{cone}\,(P_a) \subset \textbf{persp}\,(\mathrm{P}) : \Rightarrow : \mathrm{P}^\iota \in \textbf{Pl}(\mathrm{P}, \mathrm{P}')\,. \Rightarrow\,. \mathrm{P}^\iota \in \textbf{C}(\mathrm{P}, \mathrm{P}')\}$$
Translation: For all fundamental particles P, P′, P^ι, $\mathrm{P}^\iota = \{\mathrm{P}^1, \mathrm{P}^2, \mathrm{P}^3, ..\}$ it holds that if signals emitted at $P_e \in \mathrm{P}$ and absorbed $P_a \in \mathrm{P}$ are reflected at events P^ι_r contained in various particles P^ι, then if the particles P^ι belong to $\textbf{Pl}(\mathrm{P}, \mathrm{P}')$, they also belong to $\textbf{C}(\mathrm{P}, \mathrm{P}')$.
Proof: By ax.7, th.9, and instantiation of df.24.

AX.8 *Space has Three Dimensions.*
$$\forall \mathrm{P}', \mathrm{P}'', \mathrm{P}''' : \mathrm{P}', \mathrm{P}'', \mathrm{P}''' \in \textbf{Subst} \wedge \mathrm{P}'\mathrm{P}'' = \mathrm{P}''\mathrm{P}''' = \mathrm{P}'''\mathrm{P}'$$
$$\Rightarrow : \exists 2\,\mathrm{P} : \mathrm{P} \in \textbf{Subst} \wedge \mathrm{P}\mathrm{P}' = \mathrm{P}\mathrm{P}'' = \mathrm{P}\mathrm{P}''' = \mathrm{P}'\mathrm{P}''$$
Translation: For any three equidistant particles, members of the Substrate, there are two and only two more particles, members of the Substrate, that remain equidistant from the other three.
Comment: The difference between the two tetrahedra is the *origin af handedness* in 3-space.

AX.9 *Signals between AP follow tracks between FP.*
$$\forall Q'_e, Q''_a : Q'_e \preceq Q''_a : \Rightarrow : \exists \mathrm{P}', \mathrm{P}'' : \mathrm{P}', \mathrm{P}'' \in \textbf{Subst} \wedge \mathrm{P}'Q'(Q'_e) = \mathrm{P}''Q''(Q''_a) = 0$$
Translation: If the event Q'_e in Q′ caused the event Q''_a in Q″, then there were FPs P′ and P″, so that Q′ coincided with P′ at Q'_e, and Q″ coincided with P″ at Q''_a
Comment: The Substrate serves as an "aether", i.e., a "medium for the transmission of photons"; hence all "light-tracks" are describable by reference to fundamental particles.

Df.25 *Flatness of Space.*
$$\textbf{FlatSp} : \equiv : \forall \mathrm{P}^0, \mathrm{P}^1, \mathrm{P}^2, \mathrm{P}^3, \mathrm{P}^4, \mathrm{P}^5, \mathrm{P}^6 : \mathrm{P}^0, \mathrm{P}^1, \mathrm{P}^2, \mathrm{P}^3, \mathrm{P}^4, \mathrm{P}^5, \mathrm{P}^6 \in \textbf{Subst}\,. \Rightarrow\,.$$
$$\mathrm{P}^0\mathrm{P}^1 = \mathrm{P}^0\mathrm{P}^2 = \mathrm{P}^0\mathrm{P}^3 = \mathrm{P}^0\mathrm{P}^4 = \mathrm{P}^0\mathrm{P}^5 = \mathrm{P}^0\mathrm{P}^6 = \mathrm{P}^1\mathrm{P}^2 = \mathrm{P}^2\mathrm{P}^3 = \mathrm{P}^3\mathrm{P}^4 = \mathrm{P}^4\mathrm{P}^5 = \mathrm{P}^5\mathrm{P}^6 = \mathrm{P}^6\mathrm{P}^1$$
$$\Rightarrow \exists \textbf{Pl}(\mathrm{P}, \mathrm{P}') : \mathrm{P}, \mathrm{P}' \in \textbf{Subst} \wedge \mathrm{P}^0, \mathrm{P}^1, \mathrm{P}^2, \mathrm{P}^3, \mathrm{P}^4, \mathrm{P}^5, \mathrm{P}^6 \in \textbf{Pl}(\mathrm{P}, \mathrm{P}')$$
Translation: Space is flat, or Euclidean, iff for all particles $\mathrm{P}^0, \mathrm{P}^1, \mathrm{P}^2, \mathrm{P}^3, \mathrm{P}^4, \mathrm{P}^5, \mathrm{P}^6$, members of the Substrate, if $\mathrm{P}^1, \mathrm{P}^2, \mathrm{P}^3, \mathrm{P}^4, \mathrm{P}^5, \mathrm{P}^6$ form a regular hexagon with P^0 in its center, all distances between neighbouring corners being equal to their distances (radii) from the center, then there is a plane, subset of the Substrate, to which they all belong, the six corners as well as their center.

Comment: This definition leaves the issue open whether space is Euclidean or non-Euclidean. If the flat hexagon does not fit into a plane, leaving open areas, the plane (space) is *hyperbolic*. If the hexagon does not fit into a plane, due to overlapping areas, the plane (space) is *spherical*. This is explainable (even to children) by comparing flat to saddle-like and ball-shaped surfaces.

Df.26 *Relative Simultaneity of Events* (P-*simultaneity*).
$$sim_P(P'_r, P''_r) \,.\, \equiv \,.\, P, P', P'' \in \boldsymbol{Subst} \land P_e, P_a \in P \land P_e \to P'_r \to P_a \land P_e \to P''_r \to P_a$$
Translation: Two events of reflection, P'_r and P''_r, are P-simultaneous relative to observer P, iff P, P', P'' all belong to the Substrate, and P_e triggered P'_r and P''_r, which then triggered P_a.

Th.11 P-*simultaneity is Reciprocal between the Points on any Sphere centered at* P.
$$\forall P'_r, \forall P''_r : P'' \in \boldsymbol{S}(P, P') \subset \boldsymbol{Subst} \,.\, \Rightarrow \,:\, sim_P(P'_r, P''_r) \Leftrightarrow sim_P(P''_r, P'_r)$$
Translation: The event P'_r is P-simultaneous with the event P''_r iff P''_r is P-simultaneous with P'_r.
Proof: Follows from ax.3, combined with df.12 and df.26.

Th.12 P-*Simultaneity is Transitive between the Points on any Sphere centered at* P.
$$\forall P'_r, \forall P''_r, \forall P'''_r : P'', P''' \in \boldsymbol{S}(P, P') \subset \boldsymbol{Subst} \,.\, \Rightarrow \,:$$
$$sim_P(P'_r, P''_r) \land sim_P(P''_r, P'''_r) \,.\, \Rightarrow sim_P(P'_r, P'''_r)$$
Translation: If FPs P'', P''' belong to the sphere centered at P and defined by FPs P, P', then:
if P'_r is P-sim. with P''_r and P''_r is P-sim. with P'''_r, then P'''_r is P-sim. with P'_r.
Proof: Follows from th.11 and an extension of df.26 to cover three FP instead of only two.
Comment: By increasing or decreasing distances *ad libitum,* and inserting midway-particles, the transitivity of P-simultaneity can be generalized to cover all FP in the substrate!

Scholium. Combining theorems 10-12, P-simultaneity is definable for the entire Substrate. However only a finite number of the points contained in a line can be covered by our procedure. In order to ensure simultaneity for all points on a line we are therefore faced with the choice between postulating atoms of time, thus also of space, or postulating simultaneity to be absolute. We shall here prefer the second option:

AX.10 *Simultaneity is Absolute in the Substrate.*
$$\forall P, P', P'', P''' : P'', P''' \in \boldsymbol{S}(P, P') \subset \boldsymbol{Subst} \,.\, \Rightarrow \,.$$
$$\forall P'_r, \exists 1 P''_r, \exists 1 P'''_r : sim_P(P'_r, P''_r) \land sim_P(P'_r, P'''_r) \land sim_P(P''_r, P'''_r)$$
Translation: For any triplet P',P'',P''' of particles in the Substrate, and for any event on any member of this triplet, there is one and only one event on each of the two other members of the triplet that is simultaneous with the first event, as judged from the center P of the triplet.
Comment: From ax.s 9 & 10 taken together it follows that P-simultaneity is transitive between events in the entire substrate, i.e., between events contained in AP as well as in FP.

Df.27 *Zig-zag Signals.*
$$zzs(P, P') \,.\, \equiv \,.\, P_0 \to P'_0 \to P_1 \to P'_1 \to P_2 \to P'_2 \ldots$$
Translation: There was a zig-zag signal between P and P' iff P_i in P and P'_i in P' formed a chain of events $P_0, P'_0, P_1, P'_1, P_2, P'_2 \ldots$ following each other in immediate causal succession.

Df.28 *Cosmic Clocks.*
$$ccl(PP') \,.\, \equiv \,.\, \exists\, zzs(P, P') : P, P' \in \boldsymbol{Subst} \land$$

Mogens True Wegener

$$\{..., P_0, P_1, P_2, ...\} \rightharpoonup \{..., T(P_0), T(P_1), T(P_2), ...\} \subseteq \mathcal{R}$$

Translation: The cosmic clock carried by P is a Langevin clock with unit $PP' = 1/\Delta T$ defined by a zig-zag signal exchanged between P & P', P & P' being both FPs, and successive events P_i being mapped on instants $T(P_i)$ constituting a subset of the set of real numbers $\mathcal{R}$.

Th.13 *There are Cosmic Clocks.*
$$\forall P, \forall P' : P, P' \in \boldsymbol{Subst} . \Rightarrow : \exists \boldsymbol{ccl}(PP')$$
Translation: For any pair P, P' of FPs, we can construct the cosmic clock $\boldsymbol{ccl}(PP')$.
Proof: Follows immediately from ax.s 1 & 10 combined with df.28.

Df.29 *Non-Identity of Cosmic Clocks.*
$$\boldsymbol{ccl}(PP') \neq \boldsymbol{ccl}(P'P)$$
Translation: Two cosmic clocks are different iff their carriers are different.

Df.30 *Similarity of Cosmic Clocks.*
$$\boldsymbol{ccl}(PP') \simeq \boldsymbol{ccl}(P''P''') . \equiv . PP' = P''P''' = 1/\Delta T$$
Translation: Two cosmic clocks are similar iff their units are equal.

Cor.3 *Similarity of Cosmic Clocks is Reciprocal.*
$$\forall P, \forall P' : \boldsymbol{ccl}(PP') \simeq \boldsymbol{ccl}(P'P)$$
Translation: Two different cosmic clocks defined by the same pair of particles are similar.
Proof: Follows immediately from their defining units being equal.

Cor.4 *Similarity of Cosmic Clocks is Transitive.*
$$\forall P, \forall P', \forall P'', \forall P''' :$$
$$\boldsymbol{ccl}(PP') \simeq \boldsymbol{ccl}(P'P'') \wedge \boldsymbol{ccl}(P'P'') \simeq \boldsymbol{ccl}(P''P''') . \Rightarrow . \boldsymbol{ccl}(PP') \simeq \boldsymbol{ccl}(P''P''')$$
Translation: Similarity is transitive between any triple of cosmic clocks.
Proof: Follows immediately from their defining units being equal.

Df.31 *Signal-Functions.*
$$T^{P'} = \Theta^{PP'}(T^P). \equiv . T(P_i') = \Theta^{PP'}(T(P_{i-1})). \equiv . \exists \boldsymbol{zzs}(P, P') : T(P_{i-1}) \rightharpoonup T(P_i')$$
Translation: There is a signal-function from instants $T(P_{i-1})$ on the clock $\boldsymbol{ccl}(PP')$ carried by P to instants $T(P_i')$ on the clock $\boldsymbol{ccl}(P'P)$ carried by P' iff there is a zig-zag signal between P & P' together with a mapping of preceding instants $T(P_{i-1})$ in P to succeeding instants $T(P_i')$ in P'.

Df.32 *Mapping a Clock onto Itself.*
$$\boldsymbol{ccl}(PP') \rightharpoonup \boldsymbol{ccl}(PP') . \equiv . T_i^P = \Theta^{P'P}\Theta^{PP'}(T_{i-2}^P). \equiv . T(P_i) = \Theta^{P'P}\Theta^{PP'}(T(P_{i-2}))$$
Translation: There is a mapping of the clock $\boldsymbol{ccl}(PP')$ onto itself iff there is a zig-zag signal from P to P' and back together with a signal-function from P to P' mapping instants in P onto instants in P' and another signal-function from P' to P mapping instants in P' onto instants in P.

Df.33 *Congruence of Cosmic Clocks.*
$$\boldsymbol{ccl}(PP') \equiv \boldsymbol{ccl}(P'P) . \equiv . \Theta^{P'P} = \Theta^{PP'} = \Theta$$
Translation: The clock $\boldsymbol{ccl}(PP')$ of P is said to be congruent with the clock $\boldsymbol{ccl}(P'P)$ of P' iff the signal-function from P to P' is identical to the signal-function from P' to P.

Comment: Clocks are made congruent by *regraduation,* i.e., a formal adjustment of unit & zero. A method for deriving the functional square root of $\Theta^{P'P}\Theta^{PP'}$ was found by Milne & Whitrow. That congruence of cosmic clocks is reciprocal follows at once from the definition just given. However, it was also shown by Whitrow & Milne that congruence of clocks is transitive among collinear particles iff their signal-functions commute.

Th.14 *Congruent Cosmic Clocks are Similar.*
$$\forall P, \forall P' : ccl(PP') \equiv ccl(P'P) . \Rightarrow . ccl(PP') \simeq ccl(P'P)$$
Translation: If two cosmic clocks are congruent they are also similar, i.e., keep the same rate.
Comment: Similar clocks keep the same rate. Congruent clocks also agree on a common zero.

Df.34 *Atomic Clocks.*
$$acl(Q, r_a) . \equiv . Q \in Univ \wedge unit = r_a = 1/\Delta t \wedge$$
$$\{..., Q_0, Q_1, Q_2, ...\} \rightarrow \{..., t(Q_0), t(Q_1), t(Q_2), ...\} \subset \boldsymbol{Z}_\pm$$
Translation: The atomic clock carried by Q is a mechanism devised to amplify the oscillations of atoms of a specified type displaying the natural frequency $1/r_a$, with events Q_i being mapped on instants $t(Q_i)$ constituting a very fine-grained subset of the set of natural numbers $\boldsymbol{Z}_\pm$.

AX.11 *There are Atomic Clocks.*
$$\forall Q : Q \in Univ . \Rightarrow . \exists acl(Q, r_a)$$
Translation: Any observer or particle in the universe is the carrier of at least one atomic clock.
Comment: This axiom differs somewhat from the other ten by its physical character.

Df.35 *Atomic Master Clocks.*
$$mcl(P, r_a) . \equiv . acl(P, r_a) \wedge P = P_F \in Subst$$
Translation: A master clock is an atomic clock carried by a fundamental particle P_F.

Df.36 *Atomic Slave Clocks.*
$$scl(Q, r_a) . \equiv . acl(Q, r_a) \wedge Q = Q_A \notin Subst$$
Translation: A slave clock is an atomic clock carried by an accidental particle Q_A.

Df.37 *Similarity of Atomic Master Clocks.*
$$mcl(P, r_a) \simeq mcl(P', r'_a) . \equiv . r_a = r'_a = 1/\Delta t$$
Translation: Two atomic master clocks are similar (i.e., keep the same rate) iff they are both controlled by atoms of the same type displaying the same frequencies based on the same radii.
Comment: The slave clocks of accidental particles are never similar in the way defined here.

Df.38 *A Stationary Universe*
$$Stat\text{-}Univ . \equiv . \forall ccl(PP'), \exists mcl(P, r_a) : PP'/r_a \equiv constant$$
Translation: The universe is stationary iff for any cosmic clock keeping the rate $PP' = 1/\Delta T$ there is an atomic master clock with the natural unit $r_a = 1/\Delta t$, so that $\Delta t/\Delta T = const.$
Comment: This is equivalent to saying that the radii r_a of certain atoms are constant compared to some arbitrarily chosen cosmic distance PP'. Then there is only *one natural scale of time.*

Df.39 *A Dissipating Universe*
$$Diss\text{-}Univ . \equiv . \forall ccl(PP'), \exists mcl(P, r_a) : PP'/r_a \equiv increasing$$

Translation: The universe is dissipating iff for any cosmic clock with the rate $PP' = 1/\Delta T$ there is an atomic master clock of rate $r_a = 1/\Delta t$, so that the rate $\Delta t/\Delta T$ is steadily increasing. *Comment:* This is equivalent to saying that the radii of atoms are steadily shrinking as compared to some arbitrarily chosen cosmic distance. So there are *two natural scales of time.*

Df.40 *Frames Associated with Fundamental Particles .*
$$\mathbf{Frame}(P) \, . \equiv . \, \exists\,\{Q_i \,|\, P = P_F \wedge PQ_i = k_i\,r_a \wedge \mathbf{mcl}(P, r_a) \equiv \mathbf{scl}(Q, r_a)\}$$
Translation: The fundamental particle P is associated with a stationary reference frame iff P is surrounded by accidental particles Q_i at fixed distances from P as their origo, counted by a finite number k_i of standard atomic radii r_a, each of these particles being provided with atomic slave clocks kept congruent to the original master clock of P which is situated in origo of P's frame. *Comment:* If the universe is stationary, all such reference frames coincide with the Substrate. If, by contrast, the universe is dissipating, it is the main point of our axiomatization that there are no such frames, and that their artificial construction would necessitate that the involved accidental particles were constrained by external forces to remain in their positions at fixed distances from their origo, and their associated slave clocks were constrained by external forces to remain congruent to the master clock in origo. *The Substrate is the only true Frame.*

Df.41 *Frames Associated with Accidental Particles.*
$$\mathbf{frame}(Q) \, . \equiv . \, \exists\,\{R_i \,|\, Q = Q_A \wedge QR_i = k_i\,r_a \wedge \mathbf{scl}(Q, r_a) \equiv \mathbf{scl}(R, r_a)\}$$
Translation: The accidental particle Q is associated with a comoving reference frame iff Q is surrounded by accidental particles R_i keeping fixed distances to Q as origo, counted by a finite number k_i of standard atomic radii r_a, each of these particles being provided with atomic slave clocks kept congruent to the presumed master clock of Q, situated in the origo of Q's frame. *Comment:* If the universe is stationary, all such reference frames coincide with the Substrate. If, by contrast, the universe is expanding, it is our view that there are no such frames in nature, and that they can only be devised by imposing artificial constraints.

Cor.5 *Our Tempo-Spatial Geometry presupposes a Substratum.*
It is an immediate consequence of our entire approach, that our geometrical entities as well as their formal properties are only definable in a strict sense with explicit reference to the ideal of a Universal Substrate of fundamental particles and a superposed layer of accidental particles.

Scholium. An axiomatics giving priority to *tempo-spatial equidistance*, as outlined above, would seem to be of particular importance to the biological ideas of Rowlands & Hill [2012], stressing the relevance of complex structures representing a sort of regular Platonic bodies.

$$=//=$$

3.   CONCLUSION

Albeit this presentation may have loopholes, even errors, I think it allows us to affirm the conclusion of Walker [1959], that it is possible to assign clocks to all particles in the Substrate "which are not merely *congruent*, but also *equivalent* to each other", so that we henceforth have "the product structure $\mathcal{T} \times \mathcal{C}$ on the set of all events", i.e., a *timespace* in the sense of Mercier, $\mathcal{T}$ being a temporal parameter, *a cosmic time,* and $\mathcal{C}$ being the space of particles.

Non-Standard Relativity

However, Walker considered only one type of clocks, viz., those we have called cosmic. Further, Walker considered a single Substrate only, the Substrate being comparable to a unique "spray" in the sense of Schutz [1973]. We shall here follow Milne and Walker by assuming the Substrate to be singular and, indeed, unique. However, by introducing the atomic master clocks of fundamental particles we have, in contrast to Walker, also considered another type of clocks. This move is essential, since a comparison of cosmic and atomic clocks is necessary for making sense of an answer to the question whether the universe is stationary or dissipating.

According to Eddington, "the theory of the expanding cosmos is equivalent to the theory of the shrinking atom". So it is impossible to decide whether the universe is expanding relative to the fixed sizes of its contents or whether its contents are shrinking relative to the dimensions of their stationary surroundings. It is obvious that different types of clocks, cosmic and atomic, are bound to measure spatial distances in different ways. This shows that Milne was right when he insisted that *the distinction between two basic scales of time, T & t, is mandatory* to physics. Now the principle of inertia, if valid at all, can only hold good relative to one of these scales. We shall follow Milne by assuming that it holds relative to the cosmic T-scale. Consequently we may expect that the free motion of particles is exposed to spontaneous accelerations when described relative to the atomic t-scale. This might be the key to explaining gravitation.

In fact, the phenomenon of gravitation can only emerge within a dissipating universe. In such a universe the principle of Mach would finally be vindicated. However, it would hold the other way round than presumed by Mach and Einstein, as well as by their host of adherents: *by this turn gravity would be explained by inertia, instead of inertia being explained by gravity.* The Substrate, as described according to the atomic t-scale - to most people the only natural one since they don't perceive the atoms of their bodies to be shrinking - is a natural reference frame. Within this frame all fundamental particles and their associated master clocks are equivalent.

The position of an accidental particle Q is definable, relative to an arbitrary observer O, by the position of that particle $P = P_F$ with which it momentarily coincides, just as its velocity is definable by the velocity of that particle $P' = P'_F$ relative to which it is momentarily at rest. Now Q, on account of its motion relative to P, must possess a certain amount of kinetic energy. Shifting our point of view from choosing O = P to choosing O = P', this energy cannot vanish, all fundamental observers being equivalent; only it is no longer kinetic, but dynamic (potential). Thus, to Q, it is as if the gravitational potential of the whole universe were centered in P'!

In a kinematic universe no gravitational attraction holds between fundamental particles. Each single fundamental particle being a center of cosmic isotropy, all particles are equivalent. For this reason, fundamental particles are not themselves exposed to gravitational potentials. Consequently, the identical master clocks of fundamental particles count the same cosmic time. So we reject the cosmic differential field equations of Einstein, Friedmann, and Lemaître.

=//=

CHAPTER 7

CONSTRUCTIVISM IN SCIENCE
FROM POINCARÉ TO EDDINGTON AND MILNE

=//=

*Presented at the 1st Internat. Poincaré Conf., 1994,
International Academy for the Philosophy of Science.
Rev. version (2011,2021) of paper in ACERHP 1996,
'Philosophia Scientiae' volume 1, cahiers special 1.
Reprinted with kind permission from ACERHP*

=//=

SUMMARY

*The merits of Poincaré as one of the greatest
mathematicians of all times are globally acknowledged,
but the value of his conventionalist theory of science is still
greatly underestimated, and his contributions to physics
and its philosophy have unjustly fallen into oblivion.
The aim of the present paper is to stress the importance
of Poincaré to physical theory and the theory of physics by
hailing him as the principal figure in the interplay between
classical philosophy and modern cosmology.*

*Minor sections on Eddington and Milne
are added to the main article in order to show
the influence of Poincaré, albeit mostly indirect,
on two major figures of British Cosmology.*

=//=

A FEW QUOTATIONS

=//=

"Nothing of all that which has been set forth about the universe could ever have been said if we had never seen the sun or the starry heavens; but observation of day and night, of months and seasons of the year, of equinox and solstice, has produced our knowledge of numbers, which has conferred on us the notion of time and inspired us to investigate the universe; whence we have got philosophy which is the greatest boon ever bestowed on mortal man by the heavens.

The cause and purpose of vision is this: God invented it and entrusted it to us in order that we should observe the orbits of reason in the heavens and use them to correct the circuits of our thought which are akin to them, though ours be troubled and they unperturbed, so that - when we learned to know them and to compute them rightly according to nature - we could bring order to our own errant circles by imitating those of God which are perfectly regular."

Plato [22]

=//=

"Reason does not extract its laws from nature, it prescribes them to nature ... In this way, by subsuming all phenomena under its own laws, reason is the source and origin of the general order of nature ... Simple, as is the origin of this law (of reciprocal attraction), which relies only on the relationships between spherical surfaces of different radii, nevertheless its consequences are so rich, as regards the variety of their mutual consistencies and uniformities, that not only does it describe all possible trajectories of heavenly bodies by conical sections, but it does also imply relations of such a kind to obtain between these sections, that no other law of gravitation than that depending on the inverse square of the distance can be suitable to a world system."

Kant [10]

=//=

"Does the harmony the human intelligence thinks it discovers in nature exist outside of this intelligence? No, beyond doubt, a reality completely independent of the mind which conceives it, sees or feels it, is impossible. A world as exterior as that - even if it existed - would for us be forever inaccessible. But what we call objective reality is, in the last analysis, what is common to many thinking beings, and could be common to all. That common part can only be a harmony expressed by mathematical laws. It is this harmony, then, which is the sôle objective reality, the only truth we can attain. When I add that the universal harmony of the world is the source of all beauty, it will be understood what prize we should attach to the slow and difficult progress which little by little enables us to know it better."

Poincaré [24]

=//=

1. FROM KANT TO PLATO

The merits of Jules Henri Poincaré as one of the greatest mathematicians of all times are globally acknowledged. But the value of his conventionalist theory of science is still seriously underestimated, partly due to misrepresentation by historians, partly due to unfair criticism propagated by philosophers; and his great contributions to physics and its philosophy have unjustly fallen into oblivion as compared to the overwhelming fame of Albert Einstein.

In the present paper it is *my aim to stress the importance of Poincaré to physical theory and the theory of physics by hailing him as **the** principal figure in the traditional interplay between classical philosophy and modern cosmology.* As an example, I want to highlight him as the central link in a line of development connecting the main stream of European thought, as represented by Kant's *Critique of Pure Reason*, to two seemingly incompatible non-standard cosmologies: viz., that of Arthur Stanley Eddington, and that of Edward Arthur Milne.

Within the restricted frame of time and space allowed to me at this very special occasion it is of course not advisable for me to dwell at length on historical detail, neither do I feel able to do so without further study. What I want to do is to draw, with coloured brush and sweeping gesture, some very broad lines in the history of scientific ideas. These, as I see it, opens some exciting philosophical perspectives that might in the end help us to throw light on the present impasse of cosmology. But in order not to pretend too much, I shall close these introductory remarks by reminding you of the obvious fact, that science is always in need of bold new ideas. This is one of the reasons why we should not forget about its history.

The story of how Immanuel Kant was disturbed in his dogmatic slumber by the doubts of an eloquent Scotchman is well-known: finally he was forced to confront the scandal of the contemporary philosophy. A century earlier, Isaac Newton had obtained to physics its most brilliant triumph ever, yet philosophy had been unable to account for this unique achievement, let alone to disclose its legitimacy. In spite of Descartes, the ghost of Aristotelianism was still haunting philosophy. But to Kant, at least, it had become clear that the way of abstraction is blocked: true knowledge can never be obtained by the process of successive approximation. Inspired by the feat of Nicholas Koppernigk, the great innovator of medieval astronomy, Kant now set himself the task to effect a Copernican revolution in philosophy.

In order to further "the safe progress of science", Kant proposed a distinction between plain phenomena and true noumena or, as he said, between *reality-for-us* and *reality-in-itself.* Of *phenomena*, belonging to the realm of experience and stemming from the joint venture of observation and speculation or from the teamwork of sensation and reflection, we can know everything. Of *noumena,* permanently hidden, we can know nothing; the inner nature of reality transcends our inborn intellectual capabilities forever. But apparent nature, the plain surface of reality, remains transparent to our intelligence. What we must do, in order to obtain absolute and indubitable knowledge, is only to apply those conceptions which distinguish our inborn nature as thinking beings from the manifold of those sensations that are continuously caught by the network defining the structure of our natural intuition: the framework of time and space. True knowledge of apparent nature, reality-for-us, is then bound to emerge.

Knowledge of this kind, albeit occasioned by experience, gets its validity and legitimacy from another source, viz., the collaboration of reason and intuition. According to Kant, pure reason can collaborate with pure intuition ahead of any mediation of experience, and the result of this activity is pure knowledge *á priori*. Knowledge is *á priori* if it can be constructed by strictly transcendental arguments, i.e., formal arguments which are valid independently of any concrete experience. As regards the *á priori* argument given by Kant to prove the inverse square law of gravitation, it is clear that his claim - astounding as it appears - can be sustained on the assumption that gravitational forces can be described in flat 3-space: his argument is then on a par with that leading to the so-called Olbers' paradox. Though we always have to wait for such arguments to be invented, it is nevertheless interesting to speculate how Kantianism might have been received if - *per impossibile* - the Euclidean parallel axiom had eventually been proven.

To Poincaré, the failure of all the proofs given hitherto was decisive evidence against the claim of Kant that the structure of space can be demonstrated *a priori*. Although accepting the possibility of a pure intuition of space, he claimed that such space is devoid of any formal structure, thus only definable in negative terms. This brought him close to the position of Plato who frankly stated that space, "the uterus of becoming", is nothing but an imaginary container whose dreamlike existence it "is hard to believe in"; formless, and causally ineffective, it is "next to nothing". Following Descartes, extension is a substantial property and space is material; but Poincaré rejected the Cartesian aether-hypothesis, just as it had been rejected by Leibniz, and that for the very same reason: abstract space is relational, not substantial.

It remains for us to point out that the gist of Kantian *á-priorism* is not bound up to the problem of geometry and, *á fortiori*, not to the idea of an aether. Its real issue is *the active rôle of the intellect* in the reconstruction of the universe. This activity of reconstructing the universe is performed by *the transcendental subject* which may be interpreted as a kind of Platonic demiurge stripped of mythology: the *demiurge* created *cósmos* from *cháos* by applying ideas of geometry and categories of logic to a pre-existing ocean of sensible qualities; in the same way the transcendental subject produces its world by applying reason to intuition.

But the most notable difference between Kant and Plato is that Plato considered physics to be "the science of the probable", not *epistéme*, but *dóxa,* whereas Kant insisted on *epistéme,* i.e., a final and absolute knowledge of nature. At this point modern science is clearly much more close to Plato than it is to Kant. Plato dreamt of a kinship between mind and nature, between concept and reality. He assumed that harmony is inherent in the world of nature, believing that it can be discovered by human reason because reason itself is part of that harmony.

I will elaborate a little more on this idea, as expressed by Poincaré, Eddington and Milne. It turns out that the modern equivalent to the ancient idea of harmony is mathematical: group-theoretical isomorphism. This is the clue to the art of world-building in modern cosmology.

2. JULES HENRI POINCARÉ (1854-1912)

In contrast to the theme of this conference, *Science et Hypothése*, which refers to the first of the four books on the philosophy of science written by Poincaré [1902], my own reflections will primarily put focus on the second of these, viz. *La Valeur de la Science* [1905], which I take to be the most important of his books, the jewel in a quartet of precious stones.

Mogens True Wegener

My main reason for preferring this book is the primacy it gives to time ahead of space. My exposition of the philosophy of Poincaré will be strongly influenced by Giedymin [1982] who described the aim of Poincaré as follows: to examine the evolution of science and to show that progress is real, in spite of the radical changes transforming scientific theories.

Following Poincaré, *the search for truth* is the sole end worthy of science. Truth must be pursued in a spirit of righteousness, without prejudice and passion. But just as nature in itself is beautiful, so *the truth of nature is beautiful*, cf.p.6, Keats. If it were not, it would not be worth knowing, and life would not be worth living. Although there are worlds of difference between the passionate pursuit of beauty, the dispassionate search for truth, and the unselfish devotion to a higher purpose, these three cannot, and should not, be separated. As ideal values they are of the same kind, and whosoever truly loves the one cannot help loving the other two as well. The world is one. For this reason *art, science and morals* belong together.

Poincaré, being more dedicated to science, speaks primarily of *scientific truth*. In order to attain this goal, science must strive for *unity*, *simplicity*, and *objectivity*. Now experience is the only source of truth, and the ultimate arbiter of our theories, but it is we who decide how to search for truth, and when to trust the evidence; our hypotheses, our criteria and our method are our own choice and responsibility. Elaborating his own special brand of the widely accepted hypothetical deductive method, *Poincaré attempts to steer a middle course* between what he considers to be extremes: the pure *á-posteriori*, as propagated by the positivistic empiricism of his own time, and the pure *á-priori*, as advocated by the transcendental criticism of Kant.

Contrary to these extremes, he holds scientific theories to be constructions, or artefacts. The whole enterprise of natural science is *constructive*, aiming at a reconstruction of the formal relations which inhere in nature and, in this sense, science is also *descriptive*. Science never bothers about particular facts, its only concern being classes of facts; what it describes is the order or structure of facts, not their essence or substance: *disregarding substance and matter, science focusses on order and form*. Science seeks regularity in order to predict; only repeatable facts can be predicted. The only facts of relevance to science, therefore, are those that can be repeated, so the first step towards science is a preliminary classification of observables.

Scientific facts are nothing but common-sense facts expressed in the language of science. The language of exact science is an artificial one, namely the formal language of mathematics. The final outcome of scientific research is *a scientific theory*, and such theory is a harmonious *mathematical structure mapping the objective or invariant relations between observable facts*. Just as an artist selects those features of his model which perfect his picture of it, his refined sense for congruity induces him *á priori* to select precisely those facts which conform to his preconceived ideas and hypotheses of the universal harmony. But *it is a gross mistake to believe that the scientist creates his own facts: all that is manufactured in a fact is the formal language in which it is enunciated*, and it never depends on the scientist whether his prediction is fulfilled. Thus *empirical reality remains the ultimate arbiter of theoretical speculation*.

Poincaré thereby assumes a balanced position - apparently perfectly traditional - equally far away from all the excesses of nominalism or realism. It is customary to describe his position as *conventionalism*, although he did not make use of that term himself; but his particular version of conventionalism is moderate. Between the extremes, he discusses and repudiates the radical conventionalism, far from his own view, proposed by some contemporary devotees of idealism. I agree with Giedymin that the term *constructivism* may seem more appropriate.

Non-Standard Relativity

In any case it is clear that his position assumes an empirical foundation amounting to the existence of a kind of *observational invariant* beneath all theoretical conventions. This invariant reality can be known up to the structural *isomorphism* of rival theories. Changes mostly concern ontologies and metaphors but seldom affect formal structures. This enables us to make steady progress in our knowledge of reality, but our knowledge remains limited in the sense that this *reality consists in no more than what is describable by a structure of group transformations.*

Both physics and geometry study invariants under the transformation of groups. Whereas *geometry* studies the properties of ideal space, *physics* studies the temporal changes of relations obtaining between objects situated in ideal spaces. Now *the passage of time is real* whereas *space is nothing but a word* wrongly supposed to refer to a real thing or frame: *real empty space simply does not exist*, and what experience informs us of is only the relations holding between solid bodies. However, the structure of time or space is not forced upon us by nature, it is we who impose it upon nature, though not by *á priori* intuition, but because it is convenient.

With respect to space we must distinguish between: *a*) solid *bodies* whose qualities are manifested to our bodily senses; *b*) their quantitative *relations* which are measurable relative to our standards fixed by convention; and eventually; *c*) the formal *spaces* of geometry proper. To Poincaré, *geometry* is just *the formal study of groups*. It is based on premises chosen by considering their fruitfulness and appropriateness in our description of physical phenomena. As definitions, postulates, and rules of inference, its premises lead to consequences which are derived by means of exact *analysis*. But the inventiveness of mathematical construction depends on imaginative intuition; for this reason *intuition* is our best guide to fruitful *synthesis,* cf. Kant. Geometrical space is taken to be continuous by convention and, a continuum having no intrinsic metric, the concept of metrical congruence is a convention. Thus, *on the whole, the assumptions of geometry are based on convention*, their foundation being neither synthetic, nor analytic.

Experiments teach us the relations of bodies to other bodies; they tell us nothing about the relations between bodies and space, nor about the relations between different parts of space. Following Poincaré, *the only relations existing in nature are the non-metrical relations of order* which are expressible in *topology*. Dismissing the Kantian position that the geometry of space is derived by synthesis *á priori*, Poincaré upheld a kind of *á-priorism* as regards the foundation of analysis and did not interpret the axioms of arithmetic as implicit definitions of primitives. Thus he did not extend his geometrical conventionalism to arithmetics which he took to depend on strict intuition *á priori*, based on whole numbers and the principle of mathematical induction. Further, he insisted that the consistency of geometry be evaluated relative to arithmetics.

Syntactically, *a geometry is nothing but an abstract formal language*. Formal languages may treat of various objects - points, lines, planes - yet they may be identical in their structure, due to group theoretical isomorphism. According to Poincaré it is natural to consider *geometries* as linguistic *frameworks* rather than as experimental conjectures; as frameworks they cannot be put to test, of course, but this does not imply that they must remain unchangeable. The choice of a geometry to correlate experiental facts is an opportunistic affair: does it provide us with the best means to solve the central problems of physics? To Poincaré, physical reality is knowable merely up to *the observational equivalence of alternative theoretical systems* and their structural isomorphisms. The question therefore is: can we avoid falsification of our physical ideas simply by constructing a new language which is formally translatable to the old one?

According to Giedymin, Poincaré adopted a general version of the Duhem-Quine thesis: *falsification is possible only relative to systems of hypotheses expressed in a fixed language.* Therefore, instead of blaming one or more of our hypotheses in face of contrary evidence, we may blame the experimental evidence, or we may avoid falsification by changing our language. The *language* of science is not fixed forever, but may be changed in response to experience.

Changing the *lexicon* is merely a subterfuge which tends to conceal the real problems. Changing the *syntax* of the language of science goes much deeper. It is not fruitful, however, to change our language in order to avoid falsification. Sometimes we have to accept facts as final and, if we do not, we condemn science to barrenness. But, on the whole, the language of science is based on conventional decision. Poincaré also extended his conventionalism to an analysis of the measurement of time and of the principles of physics. This was the reason for applying the term *conventionalism* not only to his epistemology but to his entire philosophy of science. Nevertheless, as already intimated, the term *constructivism* may cover his position better.

According to Poincaré, geometrical space is invented to ensure consistency regarding our reasoning about bodies and their relations: *spatial positions are not real properties of bodies, and bodies do not exist in real space, we just reason as if they did; further, temporal dates are not real properties of what happens to bodies*, but signifies the sum total of relations between events related to bodies. He finally exposed the *simultaneity* of instants and the congruence of durations to a trenchant analysis: distant simultaneity, e.g., is neither a datum of observation, nor is it a consequence of the temporal continuum, it just depends on convention. Therefore *time and space are both amorphous*, meaning that *they possess no real or intrinsic metric.*

Since the enunciation of physical laws varies with the conventions adopted, and since alternative conventions modify even the natural relations between the laws, it may be doubted whether there are among these laws any that can play the rôle of a *universal invariant* which is completely independent of linguistic conventions. However, if we introduce fictitious beings possessing senses analogous to ours, admitting the basic principles of our logic, it is a plausible conjecture that their language, however different to ours, will always be capable of translation into ours; but the possibility of translation implies the existence of something that is invariant. To translate between two different languages is precisely to disclose what remains invariant. The invariant laws, then, are those relating the supposed *ordinary facts*, whereas the relations between *scientific facts* always depend on certain basic conventions.

Poincaré distinguished between *three types of hypotheses:* formal principles, inductive generalisations and realistic interpretations. *A formal principle* is always a convention; as such it is *á priori* in a relative sense, its status being similar to that of a *real definition*; cf. Leibniz. *An inductive generalization* can be viewed as an *experiental law*, thus it is subject to repeated revision, but may at a later stage be promoted to the status of principle; such a law expresses a relation between two terms, a conceptual one and a factual one. If a law is elevated to the status of principle, a third term is introduced to mediate between the first two; hence the first relation is split up into two other relations, viz., a theoretical one between two conceptual terms, and an empirical one between a conceptual and a factual term. *A realistic interpretation* of the terms is neutral if it does not affect the relations between the terms, although the terms may be changed; the same geometry, e.g., may result whether we begin with points, with lines, or with planes.

The physics of our own time is the physics of the principles, said Poincaré. Any law can be broken up into an *á priori* principle and an *á posteriori* law; for this reason the number of scientific principles has increased and still increases while a conceptual structure is taking form. But however far the partition be pushed, there will always remain laws which are in need of being tested against experience and, if not, science as we know it would be brought to an end. We cannot satisfy all conceivable principles at once in the face of evidence, and if a principle ceases to be fertile, experiment will have condemned it without contradicting it; the reason is that *if a principle cannot be refuted by any experience, it is no longer informative:* we can infer nothing from it; and, of course, it is *useless to heap up hypotheses.*

In todays physics empirical generalizations are steadily upgraded to theoretical principles. From an epistemological point of view, the difference between geometry and physics is that, whereas all *principles of geometry* are conventional, only some of the *principles of physics* are. The importance of principles in current physics seems to be growing whereas our ability to find experimental results enabling us to discriminate between theories appears to be diminishing.

It is interesting to compare this stance of Poincaré to the position of Niels Bohr, that the description of an experimental apparatus must always be given in classical terms, even in quantum mechanics. Although the precise borderline between object and subject, or between reality and apparatus, seems to be made by an arbitrary decision, a cut must certainly be made. But, in contrast to Bohr, Poincaré remained open to a revision of classical physics.

As regards the contribution of Poincaré to Special Relativity *(SR)*, my own position is in line with the judgment of Keswani [1964f.]: *Poincaré had the whole theory, including 'ondes gravifique', and had it before Einstein, the only important difference being one of emphasis.* The fundamental questions presented by the *Relativité Resteinte*, as Poincaré saw it, was this: will not the principle of relativity, as stated by Lorentz, impose upon us entirely new notions of time and space, thereby forcing us to give up our most cherished classical ideas?

But how could Poincaré describe the principle of relativity as being unfalsifiable and yet think of giving it up in face of seemingly negative experimental results? There is a very straight-forward answer to this question: on a par with Newton, he did not want to heap up hypotheses! However, he did not accept that *SR*, or *GR*, could ever force us to renounce conventionalism: even spacetime is a conventional framework that should not be confused with external reality. He saw space as amorphous, so he dismissed the intrinsic metric proposed by Einstein.

Poincaré was well acquainted with the work of Einstein, but did not credit him with the invention of *SR*. After emphasizing that the simultaneity of two events, as well as the equality of two durations, should be so defined that the natural laws may become as simple as possible, he credited Lorentz for having saved relativity by means of his ingenious idea of local time. Lorentz later, paying tribute to Poincaré for his great contributions to physics, praised him for having stated the relativistic transformations in their most convenient form, ahead of Einstein and Minkowski. Lorentz had claimed that the forces of physics should be defined so as to make them invariant to his transformations. Poincaré, accepting this claim, tried to modify Newton's law of gravitation by constructing a Lorentz-invariant action-at-a-distance theory. In the view of North [1965]: had the sympathy not been so decisively in favour of a field theory of gravity, "Poincaré's memoirs might have become a turning point in the history of the subject".

It has been noticed by Stump [1989] that the burden of conventionalism is to explain the conventionality of the fundamental principles of science in a relationist way, without relying on arguments of under-determination. So, in order to be consistent, conventionalism must explain the relational origin of both gravitation and inertia. In fact, Poincaré believed that acceleration depends only on the external relations between bodies: velocity and acceleration being on a par, both have to be relative. As he felt obliged to find a solution in terms of bodies and the forces acting upon them, he did not consider the possibility of reducing gravity to spacetime structure. According to Stump, Einstein's spacetime theory of gravitation constitutes a disproof of the possibility of a pure relationalist framework. But this view is premature, to say the least.

Thus Roxburgh & Tavakol [1975] have written an important paper displaying the close affinity between the gravitational theories of Poincaré and Milne. They see the great value of Poincaré's action-at-a-distance theory in the fact that it has led to the discovery of an entire family of consistent theories which cannot be geometrized in a Riemannian manifold, but only in the more complex framework of Finsler. The cosmological solutions for these theories are derivable by means of a generalized version of the kinematic technique invented by Milne.

3.   ARTHUR STANLEY EDDINGTON (1882-1944)

It is right to say that Kant anticipated to a remarkable extent the ideas to which we are now being impelled by the modern development of physics.   Eddington [1939]

The scientific career of Eddington was very extraordinary. He took many degrees and was appointed Plumian professor of astronomy at Cambridge University when he was only 31. He made great contributions to astrophysics for which he was deservedly famous, and he wrote a lot of books, scientific and popular, which were much acclaimed. At his heigth he enjoyed a public authority almost second to none; but the humorous style of his books, exceedingly well written, was also provocative and earned him much opposition. He exposed himself and became a favourite target of criticism for positivist philosophers armed with heavy irony, but with scant sense of humour. What specifically arose the hostile feelings of many of his scientific collegues was his insistence on the possibility of mapping the structure of the universe *á priori*.

Whittaker [1947] ascribed the following principle to Eddington: *It is possible to calculate the exact values of all pure numbers describing timeless relations between the basic constants of nature by á priori mathematical deduction from epistemological principles.*

By scientific knowledge *á priori* Eddington understood knowledge that is prior to actual measurement, but not prior to exact specification of the operational procedures of measurement. He claimed to have expressed in mathematical symbols what the physicist thinks he is doing when he is measuring things. Whittaker [ibid.] described Eddington as a modern Archimedes. Archimedes, by calculating π, the ratio of the area of a circle to its squared radius, assumed qualitative geometry in order to deduce quantitative geometry. Eddington allowed himself to use everything physical except the numerical values of constants of nature: these he claimed himself able to deduce from epistemological principles, in analogy to the deduction of π.

According to Galileo, the goal of natural science is to measure what can be measured and to make measurable what cannot yet be measured. Science focusses on the quantitative aspect of nature by effecting a reduction of quality to quantity. To say that science is based on experiment and observation performed by means of apparatus is to say that it is based on counting and on the readings of instruments. Eddington called such readings *pointer-readings*, the primary ones giving the intensities of the qualitities measured, and the secondary ones giving their spacetime location, the context of the experiment being described by pointer-readings of a tertiary order.

Pointer-readings mark the coincidence of spacetime events and are quantities produced by our operations; they are not properties of nature, but relations between object and observer. All the observable variety in the universe stems from the diversity of relations between entities; nothing in the world forces us to split it up into identical units, this is just our way of thinking. But, if we consider the intrinsic nature of the entities related, nothing is left but sameness. Eddington therefore said that there is only one kind of fundamental particle: all the rich variety of elementary particles is just a manifestation in disguise of this particular sort!

Now it is possible for a group of sensations in a mind to have the same structure as a group of sensations in another mind. It is also possible for groups of entities to display the same structure, although their true nature and their properties remain unknown to us. So we can get structural knowledge of "things outside ourselves". The recognition that all physical knowledge is structural *makes obsolete the dualism between mind and matter*, Eddington claimed.

In physics, a variety of observations can be termed structural. It was Eddington's aim to construct their fundamental theory. The result of such theory is an extensive unification between the different branches of physics. Eddington did not search for new laws, he wanted to explain those already known. It is the invariance, under different circumstances, of elementary particles with simple attributes that provides us with the ultimate numerical standards of nature.

The pure numbers of nature arise as ratios of the numbers of dimensions of certain phase spaces, and our task is to calculate the number of dimensions of such spaces, he insisted.

To solve this task Eddington invented his famous calculus of *E*-numbers, a generalized version of the even more famous Hamiltonian algebra of *quaternions*. Hamilton interpreted his quaternion algebra in Kantian terms as "the science of pure time": *arithmetic* maps the structure of our intuition of *pure time* whereas *geometry* maps the structure of our intuition of *pure space*, said Kant. Eddington, by analogy, saw his own *E*-algebra as "the science of spacetime".

According to Yolton [1960], the modern edifice of natural science is developed so far that most of the relevant data in many fields have already been collected, and so the remaining task is to unify them, formulating them in a deductive system. Yolton opines that Eddington in fact made no real *á priori* deduction of the constants of nature. The codes of empirical science are not violated by focussing upon its theoretical aspects. The laws determining the quantitative results of observation are sometimes inferable from our operational specification of the relevant observational procedures. Basic hypotheses can often be replaced by epistemological principles. New data emerging, it is often found that a different system is required. Temporary set-backs, however, cannot block the general trend of science towards unification.

Eddington has attracted quite a number of followers to join his search for a deductive explanation of the strange numerical coincidences of the universe: Paul Dirac, Pascual Jordan, Erwin Schrödinger, Hermann Bondi, Peter Landsberg, and Peter Rowlands. A small society, *ANPA*, standing for *Alternative Natural Philosophy Association*, has been formed by scientists devoted to the quest for explaining these numbers, and Eddington has recently been conferred a posthumous honorary membership. Ted Bastin & Clive Kilmister, leading members of *ANPA*, have written a series of papers on Eddington's *Fundamental Theory*, cf. Kilmister [1966], which has been followed by a study of Kilmister & Tupper [1962] on *Eddington's Statistical Theory*. Assuming the quantitative aspects of the universe to be finite and discrete, David Mc Goveran of *ANPA* has used binary algebra and computer theory to improve the combinatorial hierarchy of F. Parker-Rhodes, now accepted by *ANPA* as the common ground for further research.

Mogens True Wegener

4. EDWARD ARTHUR MILNE (1896-1950)

The so-called principle of induction .. has no content ... It is a piece of out-moded furniture, and in fundamental investigations it had better be scrapped.

There is no entity 'physical space'; there is only the abstract space chosen by the physicist as a structure in which to plot phenomena, some choices giving simpler theorems than others.

The essence of scientific freedom is the right to come to conclusions which differ from those of the majority.   Milne [1951]

As a scientist Milne never attained the fame or prominence of Eddington, nor did he become victim of so bitter and fierce an opposition; but that in itself does not make him less interesting, nor less important. His feat as a cosmologist was to construct an exceedingly simple model of a universe implying the uniform dispersion of its contents of galaxies, in accordance with a cosmological principle demanding a specific type of cosmic symmetry (isotropy), of a substratum of fundamental particles. He also showed how the superposition of a secondary set of freely moving accidental particles on this substratum gives rise to something like gravity.

The main idea of Milne was that the laws of nature can be deduced rationally by taking an individual observer's awareness of a temporal sequence of events as one's point of departure. His central hypothesis, that the laws of nature are akin to geometrical theorems, places him on a par with Eddington. But Milne did not depend on Eddington. His claim, that it is possible to deduce the inverse square law of gravitation, together with the sign of gravitation, *á priori* from some excedingly simple premises, thereby reducing gravity to inertia, may sound shocking to most scientists. What arguments did he adduce in support of this startling view?

At the dawn of history, the theorems of geometry were regarded as principles of nature. All we know of Egyptian mathematics, at least, is consistent with the view that the Egyptians regarded regularities like those summed up in the Pythagorean proposition as laws of nature. These empirical regularities were discovered by drawing up different triangles, measuring them, and experimenting with them; but as "observational laws" they were nothing but brute facts. The Greeks, deducing the laws from combinations of simpler statements, postulates, or axioms, later transformed geometry into an exact science wherein everything follows from pure theory. They thereby showed the possibility of eliminating brute facts from science.

In modern presentations of geometry, the axioms are neither brute facts nor empirical statements; instead they are definitions, i.e., minimal descriptions of what we are talking about. The theorems derived from the axioms are valid when the process of inference contains no flaw, so their truth does not depend on verification. *The tendency of all exact sciences is to pass from the Egyptian inductive phase to the Greek deductive phase*, the only question being how far this can be carried out. The extent to which it can be carried out is at the same time our measure of the degree to which we can regard the universe as being rational, Milne claimed.

The laws of geometry are derivable by pure deduction, this is evident to all; so why not assume that the laws of dynamics are likewise derivable by pure deduction? Whether rigorous deduction is possible is a question that cannot be decided *á priori*. One cannot begin by stating a programme of this kind, and then just carry it out, it derives from the *á posteriori* experience of pushing deduction as far as possible. When we introduce operational definitions, a sufficient description of the relations between real entities is provided; appeal to brute fact is unnecessary. But experience is needed is to test whether a specific world model stands up to fact.

Milne's technique of radar-signals was refined by Walker [1935f.] and Whitrow [1961], as well as by Törnebohm [1957&1963] and Schutz [1973]. A popular version of their ideas is found in the famous k-calculus of Bondi. Lucas [1973], referring to Whitrow & Milne, approves "their transcendental derivation of the Lorentz transformations" as being probably the best of all possible ways in which "bewindowed Leibnizian monads" can recover their lost harmony. According to North [1965], the independent derivations by Robertson and Walker of the *RWM* standard metric for cosmology are based on assumptions inspired by Milne.

Together with McCrea, Milne intended to revive the classical cosmology of Newton in a climate completely dominated by the ideas of Einstein, cf. North. How far this development can be taken has since been shown by Landsberg & Evans [1977]. These Newtonian world-models, which were never intented to stand up to observational test, should be carefully distinguished from the world-model of Milne as presented in his monographs of [1935 &1951].

More relevant to our present purpose, however, are the attempts of Walker [1940, 1943] and Schutz [1973] to transform the kinematic ideas of *SR* & *KR* into an exact deductive science. Their studies have in a convincing manner disclosed the unique significance of the method of radar-signals as a means to enlighten problems of modern relativity theory.

Nevertheless, it is a question if the results obtained by Walker and Schutz have benefitted sufficiently from the ingenious constructivist ideas of Poincaré. What they have obtained is a mapping of the topology of current relativity theory, and what remains to be done is to expose the conventionality inherent in the metric. This reflects, for instance, on the standard definition of simultaneity at a distance. Personally, I believe Poincaré would have welcomed an attempt to show that Einstein's dissolution of distant simultaneity is not an inevitable consequence of the topology of *SR*, but depends on his particular (conventional) choice of spacetime metric.

In cosmology, we push the deductive aspect as far as possible. But it seems as if Milne was able to make his mathematics yield more than he had put into it: his output seemed to exceed his input. If this is right, then it is no longer true that only synthetic propositions contain new knowledge; analytic propositions may likewise do that when they add to their premises "the leaven of the deductive process" (Milne). If we accept that the universe is rational, the hope of many scientists that the constants of nature can somehow be deduced may not be in vain.

Any hard-baked, or hard-boiled, scientist will traditionally hold that science and religion, whilst "on nodding terms" (Milne), have no immediate bearing on one another. On the contrary Milne held that one cannot study cosmology unless one has a "religious attitude" to the universe. Einstein, in fact, meant something similar. Cosmology assumes the rationality of the universe, but is unable to give any reason for it except the cause of nature being a rational Creator.

To Milne, the creation of the universe remains the sôle and ultimate irrationality.

=//=

Mogens True Wegener

CHAPTER 8

FUNDAMENTAL QUESTIONS

SUMMARY

In this chapter we shall allow ourselves
to consider the basic questions of physics
which are those bordering on metaphysics

= \\ = || = // =

What is Truth?
Is the World Real?
Is the World just One?
Is the Universe Infinite?
Does the Universe Expand?
Is Nature Governed by Laws?
Are Occurrences Predestined?
Is Gravitation Instantaneous?
Is Time Causally Dependent?
Does Time Involve Change?
Is Simultaneity Universal?
Is the World Contingent?
Is Nature Atemporal?
Does Time Flow?
What is Time?

\\ || //

Q1. WHAT IS TRUTH?

The question of truth is one of the great problems of philosophy. There are three different views of how truth is established, viz., *a*) by *coherence*, *b*) by *correspondence*, *c*) by *consensus*. All three views are relevant with respect to *scientific truth* which is in focus of this discussion: *a*) scientific theories *must be internally consistent*, *b*) they *must correspond to empirical facts*, *c*) they *must be accepted by a community of professional experts*. However, it should be noticed that: *a*) internal consistency is no guarantee that a theory is built on sound premisses, *b*) incompatible theories may explain many of the same facts, *c*) even a large community can be mistaken. All these reservations are particularly pertinent with respect to Einsteinian relativity ...

Not all human expressions can be ascribed a *truth-value*, even not all verbal expressions. This shows the notion of *meaning* to be far more comprehensive than that of *truth*, whence the attempt to construe a theory of meaning from a theory of truth must itself be devoid of meaning. Verbal expressions that are carriers of truth-value we normally call *statements*, or propositions. Of truth-values we ordinarily reckon two: '*true*' (1), and '*false*' (0). Operating a Boolean algebra on the system of binary numbers, we can compute the truth-value of a complex statement from those of its constituents. This logical technique can be used in the construction of logical gates which can then be implemented electronically, with (1) for '*on*' and (0) for '*off*'. By such means we are able to construct a so-called *Turing machine*, i.e., a universal computer.

Using 'p', 'q', 'r' as symbols representing simple (un-analyzed) propositions, and following Tarski, we can define *the truth of 'p'* thus: "The proposition 'p' is true *iff* (i.e.: if, and only if) p". The standard *calculus of propositions* can then be constructed on the basis of various different sets of axioms, together with rules of derivation and definitions of *wff*'s (well formed formulas). A highly simple and beautiful axiomatics is that offered by Lukasiewicz in 1924, consisting of three axioms: *L.1* $(\neg p \Rightarrow p) \Rightarrow p$, read: "If not-$p$ implies p, then p" (if undeniable, 'p' is true); *L.2* $p \Rightarrow (\neg p \Rightarrow q)$, read: "If p, then not-p implies q" (thus contradiction involves absurdity); *L.3* $(p \Rightarrow q) \Rightarrow ((q \Rightarrow r) \Rightarrow (p \Rightarrow r))$, read: "If: if p then q, then: if q then r, then, if p then r" (i.e., transfer of truth-value by classical syllogism). If, instead of analyzing propositional syntax, we want to consider the internal structure of propositions, we must turn to predicate calculus or the theory of quantification, which was extensively made use of in ch.6.

The *calculus of predicates* is a modern development of the old *subject-predicate calculus* due to Aristotle, which was found to be problematic on account of its metaphysical implications. The problem is that the subject apparently implies the necessity of referring to a thing, or entity, or substance; so the elimination of the logical subject, by way of reducing it to a description in terms of pure predicates, was meant to liberate logic from the fetters of Aristotelian ontology. For the same reason it is *very problematic to base a semantical theory on the assumption that symbols acquire a meaning by referring to things, or objects*, whatever their properties may be; *much more reasonable is it to assume that objects, as well as their properties, are constituted by way of the actions we perform on them and the operational procedures we expose them to*. In the predicate calculus, *propositions* are construed by quantifying over *variables* representing unknown objects, of which we *affirm or deny* predicates representing properties; the *quantifiers* are *operators*, universal or particular, thus giving rise to universal or particular statements, resp. Natural laws are expressible by universal propositions, boundary conditions by particular ones. One of the great issues of contemporary science is whether this cleft can be bridged. -

Mogens True Wegener

Q2. IS THE WORLD REAL?

What a silly question! Isn't the world simply a summary of what we consider to be real? Nevertheless, to answer the question isn't as easy as it seems. What do we mean by 'the world'? Is it the *sum total* of what we experience? - *now*, or in the *past*? And who are 'we', in the plural? How can we ever be sure that the world I experience is the same as the world you experience? Does anything at all remain *invariant* if the *perspective* is *shifted* from you to me, or me to you? It all seems so easy when we talk with people we know well and maybe even are very fond of, but if the identity of human persons surrounding us is put into jeopardy we are really in trouble. As long as the *communication between persons* is unperturbed we feel confident about reality, but surely, this is a very feeble foundation for a scientific conception of "the real world".

Personal perspectives on the world, when based on sense experience, are bound to differ; but we suppose that the *structure* of the real world is *invariant*, i.e., common to different people, and the task of natural science is precisely to disclose this structure, describing it by a mapping. Even if we do not agree about all details in the scientific mapping of nature, we feel convinced that, behind our sense impressions, there must be a "real world" which causes what we perceive. This is what is meant by **scientific realism**: a feeling of confidence, the belief in a "something" behind all appearances, without which the motivation to make science would probably vanish. This "something" Kant defined as "das Ding an sich" in contradistinction to "die Dinge für uns". Did Kant leave us with any hope that science shall ever succeed in disclosing "the real thing"? Not at all! Not the faintest hope whatever! The hunt is like chasing a *fata morgana* in a desert. The same view is expressed by Rowlands [2007] p.60: "There is no such thing as 'reality' ".

So, what can we hope for? Well, *maybe we are able to map the structure of phenomena, i.e., of the world as it appears to us, indeed, as it must appear,* to observation and experiment. This is what Kant would tell us: whereas it is hopeless to obtain any knowledge of the universe as-it-is-in-itself, we can at least hope to get true knowledge of the universe as-it-appears-to-us; *the reason is that the universe of appearances is not independent of the way we comprehend it.* Just as our *perceptions* are necessarily encompassed by the framework of *time* and *space*, so our *conceptions* of what is, or happens, necessarily conform to *the way we think*, indeed *must* think. *This Kantian view can be interpreted as stating a primordial version of the Anthropic Principle:* "Der Verstand schöpft seine Gesetze nicht aus der Natur, sondern schreibt sie dieser vor"; and: "So ist der Verstand der Ursprung der allgemeinen Ordnung der Natur, indem er alle Erschein-ungen unter seine eigne Gesetze fasst", [1783 §§36-38]. A similar position was held much later, on very similar premises, by Eddington: "My conclusion is that not only the laws of nature but the constants of nature can be deduced from epistemological considerations.", [1939] ch.4.

For my own part, *I would prefer to retain **truth** as a regulative idea in the sense of Kant, meaning that we should always strive for true knowledge of the real universe* behind its various appearances, even though our universe is bound to remain an unknown, maybe unknowable, X. But this moderate stance *seems to face insuperable difficulties* if we consider the finite speed with which causal effects are propagated. The observable universe then presents itself to us as a sphere of concentric shells whose age is increasing outwards with their distance from the center. So, when looking outwards, we look into *the past*, seeing the universe in a temporal perspective. But how do we pass from **world-view**, the world as it appears to us, to **world-map**, the world as it is in itself, *now*, if *absolute simultaneity* is discarded in consequence of the finite speed of the propagation of light-signals? Without this, the very notion of a *world-map* seems pointless.

Q3.   IS THE WORLD JUST ONE?

Apparently the real universe could have been many, and that in very many different ways. According to the many-worlds hypothesis of Everett and Wheeler, our world is not a universe in the proper sense of the word, but a "multiverse" consisting of an infinity of "parallel universes" co-existing "side by side", probably not in supertime, but in some timeless "super-spacetime". This "multiverse" is imagined to be *branching* "every second", or whenever something happens, the aim being to retain a *unitary description* of the entire "multiverse", represented by a unique quantum wave-function Ψ, without admitting a *collapse* of Ψ whenever an observation is made. In this way one pretends to have solved the ugly problem of "wave-function collapse".

In fact, something similar can be found in *modern tempo-modal logic* which has adopted the Leibnizian notion of "possible worlds" in order to make better sense of its semantic models. Just like the above mentioned many-worlds interpretation of quantum mechanics, tempo-modal logic assumes a *branching of possibilities*, or "worlds", *towards the future*, the present moment being the earliest branching point, and past moments connoting possibilities grasped or wasted. So the "multiverse", constituting an infinite ensemble of "possible worlds", can be visualized as a "tree of life", with "branches" pointing towards the future, its "trunk" being the actual course of past events, its "twigs" being the possible outcomes of present actions or accidental events, and each "possible world" being a linear course of future events leading forth from the present. Clearly, such "possible worlds", passing "zig-zag" from one branching point to the next one at various "angles" to each other, cannot be "parallel", but may rather be imagined as "bundles" of "world lines", or "future world courses", forking from the "trunk" at the "present present".

Now this question arises: Which status should be ascribed to the "multiverse", defined as the total ensemble of "possible worlds", or "temporal world lines"? Is it *reality*, or just *fiction*? It should here be noticed that *the many worlds interpretation* of quantum mechanics claims some wave function Ψ to determine the entire "multiverse" by comprising all possible events. The whole point of the hypothesis is that Ψ, albeit unknown, is taken to describe "everything"; hence it seems possible that we, at least "in principle", are in possession of a unitary description of the "multiverse" which enables us to comprise all possible futures in a single unified theory. Is it not legitimate, then, to say that the hypothetical wave function Ψ is at least "virtually real"? The same question turns up in relation to the "possible worlds" semantics of tempo-modal logic: are those "worlds" not, at least, "virtually real"? Here we are witnesses to a stark disagreement between *possibilists*, who are apt to answer 'yes', and *actualists*, who are disposed to say 'no'. However, if "Ockham's razor", the principle *entia non sunt multiplicanda praeter necessitatem*, is accepted, there can be no doubt about the true answer: ***Only One World*** can be real!

But there have been other attempts to support the opinion that there are many universes. Thus it has been found incumbent to buttress the "big bang" idea by several *ad hoc* hypotheses. First the idea was hailed for lending support to the cosmological principle of cosmic uniformity. Next it was realized that the resulting isotropy and homogeneity probably was a little too strong, whence the observation of small ripples in the cosmic microwave background radiation was saluted for giving rise to the inhomogeneities supposed to be needed for the galaxies to form. Then, by a second thought, it was admitted that these inhomogeneities, after all, might grow up to prevent the uniform distribution of matter in space, so that a little more mixing was needed, whence the idea of cosmic "inflation" was put forth in order to solve this "mix-master" problem. Thus we got a "theory" likening the "multiverse" to "boiling porridge", Krauss [2012] p.128!

Q4. IS THE UNIVERSE INFINITE?

In spite of such difficulties, scientific speculation has not been brought to a sobering halt. Cosmologists usually eschew *infinity* for the reason that it is not easily handled mathematically; but now the idea was seized upon in order to make sense of the so-called *Anthropic Principle*. According to this piece of wisdom, the universe is as it is, because we are its lucky inhabitants, the point being that, if it were a little different from what it presently is, we would not be alive. Since it appears unbelievable to most scientists that the universe was deliberately designed by a divine creator to the purpose of being inhabitable by living beings provided with consciousness, it seems that *infinite space* must be allowed for a *cosmic Darwinian evolution* to occur.

So the saying goes that the "multiverse", by an ultra-rapid expansion after the "big bang", was blown up to contain "bubbles", each "bubble" being a relatively independent mini-universe ruled by laws of its own, with "natural constants varying by chance", and "mostly" isotropical. With this infinity of "baby-universes" bubbling and babbling all over infinite "super-spacetime" it would be outright incredible, Smolin opines, if not some few of them were similar to our own, and one of them *is therefore, by lucky chance*, just that particular universe we inhabit. *Heureka!* How all this confusing variety of "laws" and "constants" could possibly remain compatible with a *unitary* all-embracing quantum wave function Ψ, it is probably better to forget about.

Accepting *the unison verdict of Plato, Cusanus, Leibniz, Kant & cet.: **the world is one***, we can perhaps begin to discuss seriously: are there any other ways our universe can be infinite? Already Newton confronted the question of gravitating bodies in infinite space and came to the obvious conclusion that an "island" of stars situated at rest in infinite space would be unstable; but he likewise noticed that to put an infinity of stars at permanent rest in infinite space would require a balance of forces far more precise (infinitely more) than to balance a needle on its tip. Einstein, acclaimed as the founder of modern cosmology, also wrestled with the "island world" quandary, for which he proposed his celebrated model of a static, finite, yet unbounded universe in closed, spherical space; but it was soon realized that his "needle" didn't balance either.

In his [1981], Harrison discusses all the world models compatible with general relativity: some in spherical space, an initial phase of expansion leading into a final phase of contraction, some in flat space, an initial expansion decelerating towards zero due to the brake of gravitation, and some in hyperbolic space, the expansion being accelerated towards the light speed limit. Common to all these general relativistic models is their conformity to the metric of Friedmann, their apparent viability *proving general relativity to be a technique* rather than a genuine theory. Against spatially infinite world-models, Harrison misinterprets an argument of Poincaré proving that any finite number of atoms must recur to their original configuration in space if rearranged an infinite number of times; this argument is only valid for certain types of dynamical systems. Personally, I share Harrison's disgust of the Nietzschean notion of an "eternal return", but so far he has not produced any serious objection against the concept of an infinite universe.

A unique model of an infinite universe, expanding uniformly at the speed of light from a so-called *transcendent point-event* ("white singularity"), was proposed by Milne; this model is infinite in the sense that it is at its present stage of development populated by an infinity of stars, but can nevertheless be described in two seemingly incompatible ways: as an expanding sphere, according to the t,r scale, and as eternally at rest in infinite space, according to the τ,ρ scale. My own favourite model is a modification of the Milne universe: a static sphere of finite radius, embracing an infinite number of galaxy groups in a steadily accelerating dispersion outwards.

Q5. DOES THE UNIVERSE EXPAND?

Many physicists and astronomers find it hard to accept the idea of an expanding universe. Thus most of the members of **ACG** (the *Alternative Cosmology Group, www.cosmology.info*) and **NPA** (the *Natural Philosophy Alliance, www.worldnpa.org*) unamiously reject the concept. The same applies to some of my close allies in the fight against Einsteinian dogma and myth; one might nourish the suspicion that their attitude is simply due to a shortage of imagination. Below, we shall marshal some important arguments in behalf of a dynamic cosmos.

The observations of Slipher and Hubble, that the light from distant galaxies is subject to a shift, increasing with distance, of spectral lines towards the red end of the spectrum, might have led Einstein to consider the possibility that this redshift could be due to a universal motion of recession or dispersion. However, the first to suggest a direct proportionality between distance and velocity for distant galaxies, called "Hubble's law", was the cosmologist Robertson in 1928. But already in 1917 - well before Eddington's dubious "confirmation" in 1919 of the bending of light rays predicted by Einstein and, ahead of him, by Newton as well - the astronomer de Sitter had predicted "a systematic displacement of spectral lines towards the red". His reason was that his own new static world model differed from that of Einstein on account of a strange property: if some free particles were sprinkled into the model's otherwise empty space, they would spread. In this way de Sitter got "motion without matter" where Einstein had "matter without motion". The bizarre ideas of rotating or oscillating universes we shall pass over in eloquent silence.

It has been pointed out that the idea of universal expansion should not be interpreted as an expansion outwards into pre-existing space, but as a steady increase of inter-galactic distances. But if we consider the Milne-model of 1935 this is not true, since his model can be depicted as a sphere expanding outwards into flat space, at the speed of light, from its apparent center, all the while distances between (groups of) galaxies are increasing according to the Hubble law. It may seem strange to say that the Milne universe is expanding into a "pre-existing" flat space; this would be a problem if he, with Einstein, took space to be "real", but Milne did not do that: following him, the abstract mapping of space as flat is our own free choice based on convention. In principle, *the dispersion of galaxies* should be distingusihed from *the expansion of cosmos*. This is particularly relevant when we consider my own preferred world model; with that model there is a steadily accelerated dispersion of galaxies, but no trace of a universe in expansion.

The point is that the new universe of continued creation proposed here can be compared to an instantaneous "snapshot" of the Milne universe: the cosmic sphere is no longer expanding, but has a fixed radius $\mathcal{R}_u = 2$, the limiting horizon being approximated by $\mathcal{R} = 2\,th\frac{1}{2}r$, $r \to \infty$. Nevertheless, all galaxies at rest with respect to the cosmic frame as defined by the **CMBR**, are scattered, being mutually receding in conformity with $\rho \equiv \mathcal{R}/e^\tau \equiv 2\,th\frac{1}{2}r/e^\tau \equiv const.$, ρ being a comoving coordinate assigned by the observer to each single galaxy at the instant of calibration; as opined by Pierseaux [1999], this acceleration may be caused by a so-called Poincaré pressure. The *imaginary horizon* - neither an "event horizon" nor a "particle horizon" in the usual senses - by separating a *potential infinity* of *observable* galaxies from an *actual infinity* of *unobservable* (non-existing) zero-size galaxies *is interpretable as a universal surface of demarcation enabling us to ascribe a well defined total energy to the cosmic sphere*. According to Rowlands [2007], the natural choice of this energy is *zero*. Seen this way *the cosmic sphere is a perfect black hole*, the energy gain at the center being balanced by a corresponding energy loss at the periphery. The redshift then solves the paradox of the dark sky at night ("Olbers' paradox", so-called).

Q6. IS NATURE GOVERNED BY LAWS?

The concept of physical "laws" - or "laws of nature" - may sound strange to modern ears, derived as it is from such sources as Anaximander: "Whence the origin of things thence also their demise, according to necessity, for they pay penalty to each other, atoning their trespasses in conformity with the order of time." - and Herakleitos: "The Sun will not trespass its measure, or it will be prosecuted by the servants of justice/revenge". In its more modern form, of course, it is due to Newton who claimed to have deduced from experience these *principles of motion*: 1) the law of *inertia*, 2) the law of *forced acceleration*, and 3) the law of *action and reaction*. Of these, the two first laws are contained in: $F = dp/dt = d(mv)/dt = v(dm/dt)+m(dv/dt)$, $F = 0$ giving 1), and $F \neq 0$ giving 2). To them we must add 4) the law of *universal gravitation*, $F = GMm/r^2 = ma$, with $a = GM/r^2$ as the gravitational acceleration. The first equation in F may be seen as *defining* the concept of force, the last one as instantiating an example. However, "force" still remains a hazy concept within the Newtonian world system, and it is only natural that Newton after all renounced on *explaining* gravity, satisfying himself with *describing* it.

As stated by Galileo, the *Great Book of Nature* is written in the language of mathematics, and the task of natural science is to discover the true causes (*verae causae*) behind phenomena. But according to his teacher Plato - in fact: Galileo refers to Plato in several places - this is too much to expect: *concerning natural phenomena we should content ourselves with what is the more probable instead of striving for genuine knowledge* (*episteme*: knowledge, or science). Descartes, a contemporary to Galileo, proposed the first law conservation, viz., of momentum, and intimated that, if God had created only the matter of the universe together with all its laws, the end result after a temporal process of evolution would be precisely the world as we know it. A similar view of the universe as a *mechanism*, or *clockwork*, was suggested by Leibniz who replaced the Cartesian law of conservation of momentum with that of conservation of energy, terrmed *vis viva*; however, he also warned against pushing the mechanistic viewpoint too far: as he saw it, the idea of mechanical *causality* only scratches the *surface* of phenomena.

However, in public opinion it was Hume who, by his criticism of the concept of causality interpreted as necessary connection, showed Newton's system to be a castle built in the air. The scandal of philosophy, unable to disclose a solid foundation for an exact science that had provided us with such deep insights about the working of nature, was a great challenge to Kant; his solution of the puzzle was to place the necessity in the human intellect instead of in nature. His contemporary Laplace, relying on his own wholly mechanistic explanation of the universe, confidently dismissed the traditional hypothesis of a divine Creator. However, thermodynamics raised new problems relating to its second law which were not solvable by statistical mechanics, whence some scientists advocated a phenomenological interpretation of that branch of physics. But the final break with the deterministic picture of the world occurred with quantum mechanics which supplemented a strictly deterministic wave equation by a purely statistical interpretation. One can say that quantum theory thereby reconciled the opposite views of Platon and Galileo.

According to common opinion, laws of nature are expressible by universal propositions, preferably mathematical equations relating to empirically defined objects or properties, that are necessarily valid for experiment and observation. The great question is whether such laws exist. Maybe chance is fundamental, so that all apparent laws are nothing but statistical regularities? This view is corroborated by Milne's theory: in his *KR*, the laws of gravity and electrodynamics are *statistical habits of nature induced by cosmic asymmetries*, cf. van Fraassen [1989].

Q7.   ARE OCCURRENCES PREDESTINED?

Even if the fundamental laws of quantum mechanics allow only probabilistic predictions, there are arguments for determinism - indeed fatalism - of another and very different character, a famous example being the "*master argument*" of the antique philosopher Diodoros Kronos. According to Diodoros, the following trilemma is inconsistent: 1) From the possible does not follow the impossible. 2) If something is or was the case, it will necessarily have been the case. 3) Something is possible which is not the case and never will be the case. Assuming 1) & 2) to be indispensable, he claimed the falsity of 3) to be provable; hence follows that, if something is possible, it either is the case or in some future will be the case. With plausible interpretations of 1) & 2) it can be shown formally that Diodoros was right, cf. Øhrstrøm & al. [1995], granted that not only the past, but also the future, is linear; thus it follows that the master argument can be circumvented if time is conceptualized like a "tree" with infinitely branching future possibilities. Another argument of less sophistication, but much better known, is the so-called *lazy argument* which might run somewhat like this: "Whatever I do is written in the stars (or in spacetime, e.g.) If an accident hits me it is written in the stars. If it does not hit me this is written in the stars too. So, whatever I do, I can't help it". This type of argument was countered effectively by Aristotle.

According to Aristotle (*De Interpretatione ix*): *If a man says that something will be, and another that it will not be, then it is clearly necessary that one of them must be telling the truth, i.e., if every assertion is either true or false but not both at the same time .. On this assumption nothing exists or happens by chance .. nor will anything in the future be or not be by chance .. but everything will happen of necessity .. Therefore it was always true to say of anything that has happened that it would happen. But if it was always true to say that something was or would be so, it is impossible for it not to be so or not to be going to be so .. Thus it is necessary that everything which is going to happen must happen ...* If this holds, there is no need for us to reflect on or aspire for anything, assuming that if we do this will happen and if not it will not .. But these consequences are impossible: *We know that future things do stem from our choices and actions .. Therefore, clearly, not everything is as it is, or happens as it does, of necessity ..* So, what I think is something like this: *Necessarily, either there will be or there will not be a seabattle tomorrow; but (from this it does not follow) that, necessarily, there will be a seabattle tomorrow, nor (does it follow) that, necessarily, there will not be a seabattle tomorrow.*

This argument against fatalism, showing the fallacy of distributing the necessity operator over a disjunction, can be supplemented with the following modern interpretation of quantum theoretical probability in terms of a tempo-modal concept of future-directed possibility:

Probabilities are intimately related to the future. They are a form of what might be called 'presentness of the future'. The future is present in the form of possibility. Statements regarding possibility and probability are neither 'subjective' .. nor 'objective' .. but rather 'objective in a subject related way', that is, they can only be formulated on the basis of a certain knowledge, but they are then testable by anybody who is in possession of that knowledge. In a 'monistic' philosophy of mind and matter .. this kind of 'subjectivity' is characteristic of all sorts of being. The reduction of the wave packet is nothing but a gain of information based on new knowledge. The apparition of paradox has only emerged because the meaning of the Ψ-function as being 'subject related in an objective way' was not properly acknowledged. What is then left to ponder is only a quantum theoretical description of knowledge itself.   (My translation, MTW)

C.F. von Weizsäcker [1992] p.890.

Mogens True Wegener

Q8.   IS GRAVITATION INSTANTANEOUS?

According to Rowlands, the speed of the force (not the waves) of gravity must be infinite; cf. van Flandern: *The Sun's gravity emanates from its instantaneous true position, as opposed to the direction from which its light seems to come ... No relativist has as yet, to my knowledge, devised a theory to explain how it can be that the direction of the Sun's gravitational force and the direction of photons arriving from the Sun are not parallel.* See Rowlands [2007] p.448.

This contradicts the premisses of Einstein's General Relativity (***GR***). In Rowland's view: ***GR*** is "not a theory of gravity at all". It actually provides no physical mechanism for the action of the gravitational force; much rather it just exposes the way in which gravitation is measured. Neither does it replace the Newtonian theory, rather it makes use of it by requiring that the weak field limit of the gravitational potential be the Newtonian value, which must be put in by hand. In fact, the field equations of ***GR*** merely describe the curvature of spacetime mathematically, having no real physical relation to gravity at all. The only bond between curvature and gravity is tied when the classical potential is inserted by hand into the drastically simplified equation for the radial field surrounding a point-source, the so-called Schwarzschild-solution; ibid.p.452.

But Rowland's criticism of General Relativity does not stop here; on p.478, he stresses that hard problems are associated with the idea that General Relativity is necessarily nonlinear: As a nonlinear theory it declares its own unreliability by producing unrenormalisable infinities. It is too difficult to handle for cosmology and black hole physics without drastic simplifications. It is unable to give a full description of gravity even in principle, and modifying it is of no avail. It invites the possibility of a unified field theory, but as such it is nothing but a hopeless failure. It neglects the fact that the original solution by Schwarzschild, approved by Einstein, was linear. It destroys the foundation of a series of important symmetries that would be natural without it. Taken seriously, it predicts the immediate closure of a universe filled up with zero-point energy. Defined as the first stage in an unending number of best-fit models, it excludes a unified theory. Contradicting a nonlocality corroborated by experiment, it is incompatible with quantum theory. Finally it has lead to the perverse idea that high-brow math is needed at a fundamental level!

Rowlands explains the generation of particle masses with reference to the standard model extended by the so-called Higgs-mechanism. This seems perfectly natural; fortunately the Higgs particle, having played "hide-and-seek" with the experimenters for decades, has now apparently been found, and had it not, it might have been necessary to invent a "Higgs-less" explanation. Following Rowlands, *unless one believes in some extreme version of the anthropic principle, the laws of physics, in a unified theory, must be true in all places at all epochs*, ibid.p.600. Except that Rowlands deliberately eschews cosmological models, his stance as just expressed is *very much in line with the ideas behind the model of continued creation presented in this book.* Further, his claim that the gravitational force is instantaneous agrees very well with the view of Milne that gravitation is a spontaneous consequence of local deviations from global symmetry: *in a kinematic universe there is no gravitational attraction between fundamental particles.* But Rowlands wants to derive inertia from gravity, following Mach's principle, whereas I find it more natural to derive gravity from inertia, in opposition to Mach, but in agreement with Milne. The only question is if Milne's kinematic method is applicable to a physics based on something like Rowland's remark above, cf. the so-called perfect cosmological principle of Gold & Bondi. However, it is clear that the stability of such a physics must be able to allow statistical variations of an enormous size in order to be compatible with current astronomical observations.

Non-Standard Relativity

Q9.   IS TIME CAUSALLY DEPENDENT?

There is a widespread tendency in contemporary philosophy of science to see causality as well as causal order and connectivity as being more fundamental than time and temporal order. In order to discuss this attitude properly we have to settle on a plausible definition of causality; but this is not easy, there being at least three distinct and very different theories of that concept: 1) the *probabilistic* theory, 2) the *counterfactual* theory, and 3) the theory of *covering law*.

Against all three theories it can be objected (assuming what is dubious: that they indeed pretend to explain time in terms of causality) that they presuppose what they attempt to explain. As regards *the probabilistic theory*, it must be stressed that it is hard to see how the concept of probability can be ascribed any meaning concerning events which are already present or past. Relating to *the counterfactual theory*, it must be understood that the notion of a counterfactual course of events involves the notion of past facts as being now unpreventable and irrevocable together with the very speculative imagery of past-future events that are no longer possible but which might under other circumstances have been possible at an earlier stage of development. With regard to *the covering law theory* we must distinguish between laws of classical mechanics which are reversible and deterministic, giving no clue to the difference between earlier and later, and the laws of thermodynamics, where at least the second law is itself in need of a clue as to which of the two directions of time should be viewed as leading towards increase of entropy. That the laws of quantum mechanics are themselves indifferent to temporal order is irrelevant in so far as a "wave function collapse" is requisite to the production of an observational fact.

However, there is a fourth theory of causality based on *the mark method* of Reichenbach. In his [1958] p.136, he claimed to have given a time independent definition of cause and effect, so that the relation of cause to effect can be utilized to define the relation of earlier to later. Starting with a notion of causal connection $C(E_1, E_2) \simeq C(E_2, E_1)$ which is neutral to time, he described the *causal order* of events E_1 & E_2 thus: if a small variation of E_1 to E_1^* is compatible with a small variation of E_2 to E_2^*, but not the other way round, then E_1 is cause and E_2 effect. In other words: the combinations $E_1 E_2$, $E_1^* E_2^*$, $E_1 E_2^*$ may all occur, but never this one: $E_1^* E_2$. So, apparently, we have a fool-proof definition of the temporal order in terms of the causal one. Unfortunately, it is easy to produce an exceedingly simple counter-example to this definition: drop a pea into a round bowl, vary the throw as much as you wish, then, the pea having rolled forth and back a few times, the result will always be the same: a pea in the center of the bowl! To be honest: I consider this whole enterprise to be thoroughly implausible and highly suspect, the only reasonable option being to give it up, defining causal order in terms of time instead.

Now, granted that temporal order is prior to causal order, how do we define causality? The best choice is to define it in terms of *physical laws*, and the definition I propose is this one: Consider *a well defined energetic system subject to laws determining its development* in time; granted that the *various stages* of this development display a clear and distinct *temporal order*, we shall say that *any earlier stage is causally connected to any later stage*, the earlier one being the *cause* of the later one, the later one being the *effect* of the earlier one. Of course, we have to distinguish between laws that are *deterministic* or merely *probabilistic*, in the sense of statistical mechanics or in that of quantum mechanics, just as we have to distinguish between conditions that are *necessary* to produce an effect and conditions that are *sufficient* to produce the effect. *In my view, this is the only precise definition of causality that can be given and, if the world is a system of zero energy, the definition allows us to see it as a causal chain of world-states.*

Mogens True Wegener

Q10. DOES TIME INVOLVE CHANGE?

The aim of *science* is to *describe* the present, to *predict* the future, and to *explain* the past, and the difference between science and superstition depends on its way of performing this task. Thus science *presupposes the tripartition of time in past, present and future*; nevertheless, the distinction between *determined* (past-present) and *undetermined* (future) will probably suffice. The question of *tenses* is thereby placed in the focus of our attention. I will not hesitate to brand the widespread view that tenses are fictitious as a particularly pernicious sort of superstition.

Another view, rather more plausible at a first glance, is that time is subordinate to change. Aristotle, for instance, defined time as "the number of motion with respect to before and after". Now, to Aristotle, motion meant change, and so he distinguished four different kinds of change: a) change of *substance*, b) change of *quality*, c) change of *quantity*, and d) change of *locality*, going from essential motion to superficial motion. It is ironical that the origin of modern science was conditioned by a change of attitude towards viewing spatial motion as the fundamental one; this explains why spatially extended objects were considered to be basic. But even more ironical is it that the majority of today's scientists without reflection accept Aristotle's equally outdated view that the question of existence refers to things defined as objects with changing properties. Modern logic has long ago complied with the fact that objects are conceptual constructs devoid of any inherent substantiality, confirming that only statements - not things - can be true or false. Why then do physicists - e.g., Einstein, Podolsky & Rosen - still argue in behalf of a realism that is naïve in the sense that it assumes the existence of objects (quantal systems) that after having been connected in the past by an event of interaction are no longer entangled, but independent? Why not just accept that quantum theory has put an end to the old ideal of objectivity?

The solution to this impasse is simple and natural. Don't ask what is! Ask what happens! What happens we call *events* and events, present or past, are *facts* whether perceived or not. Stricly speaking, only statements - specific linguistic expressions - can be bearers of truth-value; this seems to involve something like human consciousness, so we are at the point *where matter meets mind*, and this is why the Copenhagen interpretation of quantum theory, claiming that physics does not care about *reality* but about our *knowledge* of reality, was felt so provocative. We already opened the possibility of *facts unperceived* by human observers; maybe we can also mitigate Bohr's claim so as to accept *knowledge* which is *expressible* but not actually *expressed*, for the moment ignoring that this comes close to accepting that mind is a potentiality of matter. Noticing, that "the first object, to which such a theory (i.e., 'abstract quantum theory') is related, is not a *thing*, but a *stream*", Weizsäcker [1985 p.363], further, that "in the (concrete) quantum theory, the spatiality of objects is only a derived / secondary property" [ibid. p.391], and finally, that "if the quantum theory is taken seriously from the mathematical point of view then, stricly, there are no separate objects, but an (entangled) whole", [1992 p.329] - then it appears natural to *conceive of reality as a temporal flow, or stream, broken up by the tripartition of time* into *present events* which are *just now* made actual, *irreversible facts* which are *inevitably past*, and *future possibilities* which *may or may not be realized*. From this we construe our objects, *and for this reason the change of temporal modalities is primary to any other kind of change.* Weizsäcker has axiomatized quantum theory and special relativity in terms of temporal logic. Of course, the bare change of tense operators as applied to statements is in itself vacuous if the statements themselves are empty and nothing whatever is true, meaning truth does not "exist". This would be the case if "facts" were reversible. We shall ignore this weird speculation.

Q11. IS SIMULTANEITY UNIVERSAL?

A1.   *When an event X is happening, another event Y either has happened or not happened - 'having happened' is not the kind of property that can attach to an event from one point of view but not from another. On the contrary, it's something like existing; in fact to ask what has happened is a way of asking what exists, and you can't have a thing existing from one point of view but not existing from another, although of course its existence may be known to one person or in one region, without being known to or in another. So it seems to me that there's a strong case for just digging our heels in here and saying that, relativity or no relativity, if I say I saw a certain flash before you, and you say you saw it first, one of us is just wrong - or misled it may be, by the effect of speed on his instruments - even if there is just no physical means whatever of deciding which of us it is. To put the same point another way, we may say that the theory of relativity isn't about real space and time, in which the earlier-later relation is defined in terms of pastness, presentness, and futurity; the 'time' which enters into the so-called spacetime of relativity theory isn't this, but is just part of an artificial framework which the scientists have constructed to link together observed facts in the simplest way possible, and from which those things which are systematically concealed from us are quite reasonably left out. This sort of thing has happened before .. When .. the differential calculus was first invented, its practitioners used to talk a mixture of excellent mathematics and philosophical nonsense, and at the time the nonsense was exposed for what it was by the philosopher Berkeley, in a pamphlet entitled 'A Defence of Free Thinking in Mathematics' .. The mathematicians saw in the end that Berkeley was right, though it took them about a century and a half to come round to it. They came round to it when they became occupied with problems which they could solve only by being accurate on the points where Berkeley had shown them to be loose; then they stopped thinking of the things he had to say as just a reactionary bishop's niggling, and began to say them themselves. Well, it may be that some day the mathematical physicists will want a sound logic of time and tenses; and meanwhile the logician had best go ahead and construct it, and abide his time.*

A.N. Prior, founder of tense logic; cf. Wegener [1999], W. & al. [1996], Lucas [1999].

A2.   Interviewer: *Bell's inequality, as I understand it, is rooted in two assumptions: the first is what we might call objective reality, the reality of the external world, independent of our observations; the second is locality, or non-separability, or no faster-than-light signalling. Aspect's experiment* (indicates that one has to choose. Which one would you stick to?)

John Bell: *I think it's a deep dilemma, and the solution of it will not be trivial. It will require a substantial change in the way we look at things. But I would say that the cheapest resolution is something like going back to relativity as it was before Einstein, when people like Lorentz and Poincaré thought that there was an aether - a preferred frame of reference. -*

Interview reprinted in Brown & Davies, eds.: *The Ghost in the Atom*, Cambridge 1987.

A3.   *Ein systematischer Aufbau (der Physik) würde verlangen dass zuerst die vollständige Logik zeitlicher Aussagen entwickelt und auf sie dann erst die physikalischen Theorie gegründet wurde .. Die These dieses Buchs ist, dass eine Logik zeitlicher Aussagen fundamental selbst für die Begründung der klassischen Logik sein sollte; dass diese zeitliche Logik in den Ausdrucksweisen der Umgangssprache, vielleicht am deutlichsten in den indogermanischen Sprachen, schon implicite enthalten ist; dass die Quantenlogik eine spezielle Fassung diese zeitlichen Logik ist; und dass insofern die Quantentheorie nur der Anlass war, der uns zu dieser logischen Reflexion veranlasst hat.*   C.F. von Weizsäcker: *Aufbau der Physik* [1985] pp.52&313.

Q12. IS THE WORLD CONTINGENT?

For centuries, if not millennia, it has been the aim of philosophers and physicists to invent a theory of the cosmos presenting it as a self-explaining mechanism, cause of its own existence. It being impossible to devise a *perpetuum mobile* from a particular isolated energetic system, maybe one could construct the whole universe as such? Isn't the universe itself just *causa sui*? In that case all the divine prerogatives could be transferred from God to his supposed creation: nature itself could be considered the only God, as Spinoza, Hawking, and Krauss, would have it. One scientist who, in my view, has made one of the most promising attempts in this direction, is P. Rowlands [2007]. Without knowing his attitude to metaphysical issues precisely, I have no doubt that his ambition to construct an unique "TOE" (theory of everything) is very high.

According to Rowlands (p.2f.), "we cannot devise a unified theory simply by combining quantum mechanics and general relativity in a new mathematical superstructure", such attempts being doomed to fail because partial theories are not unified by combining them but by deriving them from a common origin: thus *zero* must be the point of *departure* as well as that of *arrival*. Only the notion of *nil*, or *nothing*, split up into *duality*, is radical enough to explain *everything*. From the point of view of physics (p.84f.), "the Dirac nilpotent equation would seem to be a perfect way of producing something from nothing", since it incorporates all groups of interest; and the conservation laws implied by $(\mathbf{k}E + ii\mathbf{p} + ijm)(\mathbf{k}E + ii\mathbf{p} + ijm) = 0$, by including mass-energy and the three kinds of charge, determine the full behaviour of all physical systems. Basing our mathematics not on the integers, but on an immediate zero totality, we may produce "a mathematical structure .. avoiding the incompleteness indicated by Gödel's theorem."

Elaborating on this (p.556f.), Rowlands proposes to start with *one symbol* representing 'nothing', and *two basic rules* (duals of a single rule): 1) *create*, a process adding new symbols, and 2) *conserve*, a process examining the effects of any new symbol on those already existing, to ensure a zero sum again. He furthermore points out that *a nilpotent universal computational rewrite system (NUCRS)*, working on an infinite alphabet that defines the semantics of quantum mechanics in terms of a universal grammar, *may suffice to determine the structure of cosmos, the genetic code, the human brain, and human language.* The *NUCRS* may thereby enable us to establish an *Evolutionary Anthropic Semantic Principle* that can describe the rules by which a sentient being is able to comprehend *Nature's Own Rules*. So Rowlands suggests the method of a "bootstrap" to perform the ultimate trick: *Ouroboros*, a snake eating itself from the tail.

I deeply admire the daringly intrepid and exceedingly original construction of Rowlands. If anyone should ever succeed in mapping the invariant rules and numerical relations of nature, it would probably be him ahead of Penrose, Hawking, Barbour, Smolin, Isham, or whoever else. But how shall we assess his claim that *NUCRS* can avoid the incompleteness theorem of Gödel? The prospect of a *closed physical system*, complete with fully integrated syntax and semantics, containing a unified description and explanation of *both mind and matter*, is not very bright:

"We may note here that it *is* possible to construct a calculus rich enough in its symbolism for the statement within itself of its .. own *syntax* .. (but *not* of) its *semantics* .. It cannot be said within any system .. that the system is *complete* .. i.e., its unproven theses and rules suffice to prove all theses (that) are true for all interpretations of their variables." - A.N. Prior [1962] p.70. The proof, Prior [1962], cf. ref. 136, is simple and particularly adaptable to Rowland's *NUCRS*.

Even if Rowlands succeeded in mapping all the invariant laws and pure numbers of all possible worlds, **the abyss between possibility and factuality** *would not be bridged ...*

Q13. IS NATURE ATEMPORAL?

Einstein is reported to have said often that the problem of the 'now' worried him seriously. The problem is that physics cannot mark out an instant as being different from another instant, and this holds even if we disregard the 'spatialization' of time which was a direct consequence of the spacetime formulation of special relativity, although this naturally accentuated the problem. Physical 'time' is measured by clocks counting, by the ordinal numbers, the recurrence of events that are regulated by cyclic processes, by increase of entropy, or by the disintegration of atoms. Attempting to disclose the laws governing the causal chains between dated events, physicists ordinarily presuppose that such counting of 'time' is indifferent to their choice of 'temporal' zero. However, the supposed 'homogeneity of temporal intervals' may be valid for the master clocks of *fundamental observers* without being valid for the slave clocks of *accidental observers*.

Now many modern physicists, especially those influenced by the reasoning of Einstein, are inclined to regard physical 'time' as an *illusion* in the sense that it cannot be ascribed a fixed direction; whence follows that the usual notion of 'temporal flow' must be even more deceptive. This view is supported by a famous (notorious) argument of McTaggart (repeated by Mellor) who ingeniously distinguished between the *absolutist A-concepts of past/present/future* and the *relationist B-concepts of before/during/after*. A deep logical chasm has ever since separated the *A-theorists*, who insist to explain the *B-series* in terms of the *A-series*, from the *B-theorists*, who attempt to interpret the *A-series* in terms of the *B-series*. Today it is a commonplace to distinguish 'tensers' from 'detensers'; but it was A.N. Prior, the founder of modern tense logic, who first gave logical import to this distinction. According to Prior, *all real existence is present*, and *only present existence is real*, the past being no longer real and the future being not yet real, just as *facts* are *true statements*, and statements, if true, are *true now*, i.e., when said or read.

As it is, *A-theorists* or 'tensers' (like Prior, e.g.) would attempt to reduce talk of instants to tensed propositions, whereas *B-theorists* or 'detensers' (like Quine, e.g.) would attempt to reduce tenses to predicates of existing instants. A sort of "half-way house" in between is taken up by 'neutralists' who prefer to treat these two positions on a par. Among the A-theorists we can further distinguish 'moderates' from 'radicals': the former would insist on utilizing modal primitives together with tenses, whereas the latter would follow Prior in his attempt to define modalities by means of tenses. Taken together, all these distinctions give rise to *four grades of tense-logical involvement*, Prior [2003], ch.xi. The system *W*, proposed in **Q15**, by extending the tense-logical system for future branching time with Peircean definitions of temporal modalities and adapting it to deal with the problem of *non-statability*, goes full way to the fourth grade.

Given some present fact, what are we able to infer with respect to its past and its future? It is a fact that you are just now reading a chapter from my book on *Non-Standard Relativity*. From this fact you can infer not only that it will always have been the case that you were reading in that book, but that it is now inevitable/unpreventable, that it will have been the case. However, you cannot infer that it was always the case that you would once read in this book, merely that since you learned reading it was always possible that you might once read in a book; but even that you could not infer if you belonged to an age before the art of writing was known. Our logic thereby makes sense of **the flow of time** by separating the now from past and future: what belongs to the past is no more possible, what belongs to the future is not yet realized, but, in pace with possibilities being annihilated, new factual truths are being created just now!

Q14. DOES TIME FLOW?

The aim and purpose of *tense logic* is to systematize reasoning with tensed propositions. In order to do so properly we must distinguish between two types of statements:

1) *temporally definite statements* - i.e., sentences with invariant truth-value

2) *temporally indefinite statements* - i.e., sentences with variable truth-value

Against this distinction it has been objected that statements of the second kind are not proper propositions, but propositional functions left undetermined due to their lack of dating. But that objection can be dismissed as soon as we give attention to their context.

Tense logic, or *the logic of change*, is relevant when we study statements in their natural context which is a context of temporal change. What we perceive is reality-in-change and, just as reality itself is emerging and expiring, thus our language, in order to represent this perpetual change, must reflect it in the successive origin and closure of the truth of its assertions. The stuff of tense logic consists mainly of temporally indefinite statements, the definite statements being those which are omnitemporal, those which mark an absolute beginning or an absolute ceasing, and those that are unique in the sense that they are true *now*, but neither true in the past, nor in the future. With tense logic the verb, or copula, can no longer be interpreted as timeless, but should always be understood as referring to the present: it is *now* the case that so-and-so.

It is my aim to sketch a new system *W* of tense logic which is *indeterministic* not merely in the sense that it permits possibles to branch towards the future, but also in the sense that it, more radically than standard tense logic, discards the idea of timeless truth by implying truth to emerge in time along with reality. Truth is nevertheless assumed to be eternal in the sense that, once established, it can never be annulled or suspended but is valid henceforth, i.e., in all future. I shall see it as a virtue of my system *W* if it succeeds in reproducing the richest variety of linguistic forms by the simplest possible expense of axioms. The system will display features derived from Aristotle, Diodoros, Aquinas, Leibniz, Kierkegaard, Peirce, Kripke, and Prior.

K_t & K_b are two very simple tense-logical systems of which soundness and completeness are provable with respect to a Leibnizian *possible-worlds* semantics, as demonstrated by Kripke. But, with K_b, time acquires a direction so that we can speak of *the arrow of time*, and for this reason alone it is natural to give priority to K_b, ahead of K_t. K_b is characterized by a successive loss of possibility. The actualization of only one among an infinity of future possibilities means that most of the conceivable futures are successively eliminated. Hence what was possible in the past may now be excluded. But, making use of Prior's concept of statability, we shall claim that this perpetual loss of possibility is compensated by a steady increase in the sum of statable truth. This corroborates the view that *the passage of time* is mind-independent in an important way.

Exemplifying the statability of a proposition p by the tautology $p \Rightarrow p$, we insist that the sum of statable truth is steadily increasing due to the fact that assertions which were not hitherto statable are becoming statable in the course of time. Being now statable, we shall assume that they henceforth remain statable, so that propositions feigning departed individuals to be present are just false. Granted this, we shall claim that what is true now will inevitably have been true. By contrast it is uncertain whether what is now statable was always statable, so often we cannot know if what is true now was always going to be true. Our system *W* thus makes a difference between past and future in the sense that the continued loss of possibilities is compensated by a successive gain of statable truth. In this sense we can speak of *a creation of truth*.

Q15. WHAT IS TIME?

The System W
AXIOMS FOR TEMPO-MODAL LOGIC

PRELIMINARIES

1. *All atomic propositions π are well formed formulae, wff.*
2. *The set W of atomic propositions contains an unique constant ω called 'the world', together with a subset of abstract propositions τ termed 'instants', 'times' or 'dates'.*
3. *All instant-propositions τ are different and distinguishable by their indices: $\tau_i \neq \tau_k$.*
4. *In a certain way the constant ω may serve to characterize successive 'nows', cf. below.*
5. *If α and β are wff, then $\neg\alpha$, $\alpha \Rightarrow even\beta$, $N\alpha$, $H\alpha$, are all wff.*
6.
For '$\neg\alpha$'	*read: 'not α' or 'it is not the case that α'*
For '$\alpha \Rightarrow \beta$'	*read: 'if α (is the case), then β (is the case)'*
For '$N\alpha$'	*read: 'henceforth α' or 'α will always obtain'*
For '$H\alpha$'	*read: 'hitherto α' or 'α did always obtain'*
7. *All the formulas above with all their combinations, and no other, are wff.*

DEFINITIONS & RULES

$df \vee$	$\alpha \vee \beta \equiv \neg\alpha \Rightarrow \beta$	read: '(either) α or β' $\equiv$ 'if not α, then β'
$df \wedge$	$\alpha \wedge \beta \equiv \neg(\alpha \Rightarrow \neg\beta)$	read: '(both) α and β' $\equiv$ 'not: if α, then not β'
$df \Leftrightarrow$	$(\alpha \Leftrightarrow \beta) \equiv ((\alpha \Rightarrow \beta) \wedge (\beta \Rightarrow \alpha))$	'α iff β' $\equiv$ 'if α then β, and if β then α'
$df P$	$P\alpha \equiv \neg H\neg\alpha$	read: 'past α' $\equiv$ 'not hitherto not α'
$df M$	$M\alpha \equiv \neg N\neg\alpha$	read: 'maybe α' $\equiv$ 'not henceforth not α'
$df L$	$L\alpha \equiv HN\alpha$	read: 'forever α' $\equiv$ 'α in all past future' $\equiv$ 'necessarily α'
$df K$	$K\alpha \equiv \neg L\neg\alpha \equiv PM\alpha$	read: 'once α' $\equiv$ 'conceivably α' $\equiv$ 'not necessarily not α'
$df T_{\tau_i}$	$T_{\tau_i}\alpha \equiv (\tau_i \wedge \alpha)$	read: 'α is true at τ_i' $\equiv$ 'α obtains at τ_i' $\equiv$ ' α at τ_i'
$df P_{\tau_i}$	$P_{\tau_i}\alpha \equiv P(\tau_i \wedge \alpha)$	read: 'α was true at τ_i' $\equiv$ 'α did obtain at τ_i' $\equiv$ 'past α at τ_i'
$df M_{\tau_i}$	$M_{\tau_i}\alpha \equiv M(\tau_i \wedge \alpha)$	read: 'possibly α at τ_i' $\equiv$ 'α may obtain at τ_i' $\equiv$ 'maybe α at τ_i'
$df K_{\tau_i}$	$K_{\tau_i}\alpha \equiv K(\tau_i \wedge \alpha)$	read: 'conceivably α at τ_i' $\equiv$ 'α might obtain at τ_i'
$df F_{\tau_i}$	$F_{\tau_i}\alpha \equiv \{M\tau_i \wedge N(\tau_i \Rightarrow \alpha)\}$	("the now unpreventable future")
	read:	'inevitably α at τ_i' $\equiv$ 'maybe τ_i and henceforth, if τ_i then α'
$df D_{\tau_i}$	$D_{\tau_i}\alpha \equiv \{K\tau_i \wedge L(\tau_i \Rightarrow \alpha)\}$	("the forever predestined future")
	read:	'necessarily α at τ_i' $\equiv$ 'once τ_i and necessarily, if τ_i then α'
$df <$	$\tau_i < \tau_k \equiv L(\tau_i \Rightarrow M\tau_k)$	
	read:	'τ_i before τ_k' $\equiv$ 'necessarily, if τ_i then $M\tau_k$'
RN	$\vdash \alpha \rightarrow \vdash N\alpha$	if α is a thesis then $N\alpha$ is also a thesis
RH	$\vdash \alpha \rightarrow \vdash H\alpha$	if α is a thesis then $H\alpha$ is also a thesis
		provided that $\vdash H(\pi \Rightarrow \pi)$ for all π in α (all π were always statable)
MP	$\vdash \alpha \ \& \ \vdash (\alpha \Rightarrow \beta) \rightarrow \vdash \beta$	if α and $(\alpha \Rightarrow \beta)$ are theses, then β is a thesis
RS	rule of substitution	the general rule allowing the substitution of equivalents
	proviso	instant-propositions, being unique, are not replaceable

AXIOMS FOR PC (The Propositional Calculus - Łukasiewicz)

$P1$	$(\neg\alpha \Rightarrow \alpha) \Rightarrow \alpha$	read:	'if, if not α then α, then α'
$P2$	$\alpha \Rightarrow (\neg\alpha \Rightarrow \beta)$	read:	'if α, then: if not α, then β'
$P3$	$(\alpha \Rightarrow \beta) \Rightarrow \{(\beta \Rightarrow \gamma) \Rightarrow (\alpha \Rightarrow \gamma)\}$		'if, if α then β, then: if, if β then γ, then, if α then γ'

AXIOMS FOR THE SYSTEM K_b (*Future Branching Possibility - Kripke, Prior*)

A1 $\alpha \Rightarrow NP\alpha$ [*A1* entails $MH\alpha \Rightarrow \alpha$]
read: 'if α, then inevitably past α'

A2 $\alpha \Rightarrow HM\alpha$ *if* $\vdash H(\pi \Rightarrow \pi)$ *for all π in α*   [*A2* entails $PN\alpha \Rightarrow \alpha$ with proviso]
read: 'if α, then hitherto maybe α, granted that all π in α were always statable'

A3 $H(\alpha \Rightarrow \beta) \Rightarrow (H\alpha \Rightarrow H\beta)$ [*A3* claims distributivity of H]
read: 'if hitherto: if α then β, then: if hitherto α then hitherto β'

A4 $N(\alpha \Rightarrow \beta) \Rightarrow (N\alpha \Rightarrow N\beta)$ [*A4* claims distributivity of N]
read: 'if henceforth: if α then β, then: if henceforth α then henceforth β'

A5 $MP\alpha \Rightarrow (\alpha \lor M\alpha \lor P\alpha)$ [*A5* entails linearity of the past]
read: 'iff maybe past α, then α or maybe α or past α'

A6 $N\alpha \Leftrightarrow NN\alpha$ [*A6* entails transitivity and density of N]
read: 'iff henceforth α, then henceforth henceforth α'

A7 $H\alpha \Leftrightarrow HH\alpha$ [*A7* would be provable with unconditioned *RH*]
read: 'iff hitherto α, then hitherto hitherto α'

A8 $N\alpha \Rightarrow M\alpha$ [*A8* claims that inevitability implies possibility]
read: 'if henceforth always α, then maybe α'

AXIOMS FOR THE SYSTEM S5 (*Omni-Temporal Necessity - Leibniz, Lewis*)

L1 $L\alpha \Rightarrow \alpha$ [In $\mathcal{W}$, *L1* is derivable from *dfL, PC*1-3, *A2, A6*]
read: 'if forever α, then α'

L2 $L(\alpha \Rightarrow \beta) \Rightarrow (L\alpha \Rightarrow L\beta)$ [In $\mathcal{W}$, *L2* is derivable from *dfL, PC*1-3, *A3, A4*]
read: 'if forever: α implies β, then forever α implies forever β'

L3 $KL\alpha \Rightarrow L\alpha$ [*L3* is a basic axiom characterizing the system *S5*]
read: 'if only it might be that forever α, then forever α'

AXIOMS FOR TEMPORAL INSTANTS (*Dates - Wegener*)

T1 $\tau_i \Rightarrow \neg M\tau_i$ *'instant-propositions are unrepeatable'*
T2 $K\tau_i \Rightarrow (\tau_i \lor M\tau_i \lor P\tau_i)$ *'the order of instant-propositions is linear'*
Cor $K\tau_i \Rightarrow L(\tau_i \lor M\tau_i \lor P\tau_i)$ *'instant-propositions are necessarily statable'*

AXIOMS FOR UNIVERSAL TRUTH (*The Present - Meredith*)

N1 ω *'the world is present'*
N2 $L\omega \Rightarrow \alpha$ *'the world is contingent'*
N3 $\alpha \Rightarrow L(\omega \Rightarrow \alpha)$ *'the world is universal truth', or*
                        'the world necessarily implies everything true just now'

For a fuller discussion of the system $\mathcal{W}$, cf.: **www.m-t-w.me**, *'A Logic of Creation'*.

*

*Oh source of grace who granted me the courage
to look so steadfast on thy blaze eternal
that all my power of vision was exhausted!
Within thy depths I clearly saw collected
all leaves that in the universe are scattered
bound up with love as in a single volume!*

Dante Alighieri: *The Divine Comedy, canto xxxiii 82f.*

*

Non-Standard Relativity

9. COMPACT CONCLUSION (rev. 2020)

df. = definition, pr. = proposition

df. 01	The Universe (U): the hidden reality confronting us in our observations and experiments.
df. 02	A world-model: a consistent formal description of U which is testable against experience.
df. 03	The universe (u): the model chosen by us for consideration among all possible world-models.
df. 04	The substratum (S): a particle-tied aether related to **c**osmic **m**icrowave **b**ackground **r**adiation.
df. 05	Fundamental particles (FP): material objects / human observers at rest in S (relative to CMBR).
df. 06	Accidental particles (AP): material objects / human observers moving in S (relative to CMBR).
df. 07	Atomic clocks: precision clocks constructed by using atoms of the same type as "Zeit-Gebers".
df. 08	Cosmic time T: the time read off properly synchronized atomic master-clocks associated to FP. (if the universe originated in a "big bang" singularity, this would serve as a common time zero).
df. 09	Absolute simultaneity: the non-locality of a quantum event apt to define a distinct world-state.
df. 10	The differential light speed: the co-efficient c between infinitesimal elements of space and time.
df. 11	The distance to and fro an object: the integral light speed times the difference between the two clock-readings marking the emission and the reception of a radar-signal reflected by the object.
df. 12	The reflection-instant of a signal: an instant in the open interval between emission and reception.
df. 13	Reference frame: a feigned spatial grid of clocks keeping fixed distances to an observer in origo.
df. 14	Proper time: the time shown by the atomic master-clock conjoined to some object, or observer.
df. 15	Frame time: the time shown by atomic slave-clocks at rest in the spatial frame of an observer.
df. 16	Einstein-convention: the standard rule for synchronizing slave-clocks at rest in a spatial frame.
df. 17	Milne-regraduation of cosmic time: transmutes the map of an expanding universe with atoms of fixed size into the map of a stationary universe of shrinking atoms & vice versa.

pr.01	All experience testifies that Time Flows, the Present separating the Past from the Future.
pr.02	The Now is the moment of Becoming when Reality emerges, annihilating empty possibilities.
pr.03	The Universe (U) must therefore be subject to the Absolute Simultaneity of a Cosmic Time T.
pr.04	Atoms of the same type, if exposed to similar conditions, partake in the same Cosmic Rhythm.
pr.05	Only a universe (u) with an infinite substratum of fundamental particles sustains such a rhythm.
pr.06	In order for there to be invariant laws, galaxies must be dissipating relative to the sizes of atoms.
pr.07	The only relevant universes are those with a structure determined by a substratum in dissipation.
pr.08	The crucial property of such universes is not homogeneity but the symmetry of Cosmic Isotropy.
pr.09	The substratum S of fundamental particles is covered by local layers of accidental particles.
pr.10	Fundamental particles, FP, are at rest in S, relative to CMBR. Accidental particles, AP, are not.
pr.11	The clocks of fundamental particles count Universal Time, but those of accidental ones do not.
pr.12	A fictitious spatial standard frame can be ascribed to any object, whether at rest or in motion.
pr.13	But it can never be supplied with identical clocks except for infinitesimal elements of the grid.
pr.14	At any time T, only one fundamental particle can be at rest in a frame, being its natural Origo.
pr.15	Any AP, say Q, refers at any instant T to two FPs: passing by FP1, it is at rest relative to FP2.
pr.16	The clock of Q is retarded relative to that of FP1 due to motion, to that of FP2 due to gravity.
pr.17	The motional energy of Q relative to FP1 equals the gravitational energy of Q relative to FP2.
pr.18	The standard dilatation of Q-time t is $1/\sqrt{1-v^2}$ relative to FP1 and $1/\sqrt{1+2\varphi}$ relative to FP2.
pr.19	Hence it is not the master clock of an FP crossing the standard frame of another FP that is slow.
pr.20	Rather the slave clocks of that frame are slow relative to the master clock of any FP showing T.
pr.21	Gravitation is a local, direct, and instantaneous, consequence of the universal dissipation of FP.
pr.22	The Einstein-Friedmann-Lemaître eqs. are null and void: gravitation is not universal, only local.
pr.23	Flat 3-space is not needed in order to ensure that the gravitational energy of FP sum up to zero.
pr.24	There is no evidence for "universal inflation", nor for the existence of "dark" matter or energy.

Mogens True Wegener

10. SYMBOLIC SURVEY (rev. 2020)

Postulate of Universal Time: $T = invar.$

differential light speed: $c \equiv \frac{space\ element}{time\ element} \equiv 1$

proper time on observers' clock: $\tau_1 \leq \tau \leq \tau_3$

one way light speeds: $\tau \equiv \tau_3 - r/c_\leftarrow \equiv \tau_1 + r/c_\rightarrow$

universal two way light speed: $\frac{1}{2}(\frac{1}{c_\leftarrow} + \frac{1}{c_\rightarrow}) \equiv \frac{1}{c} \equiv 1$

standard frame time coordinate: $t \equiv \frac{1}{2}(\tau_3 + \tau_1)$

standard frame space coordinate: $r \equiv \frac{1}{2}(\tau_3 - \tau_1)$

standard red shift: $1 + z(t) = e^{(\tau_3 - t)/t_o} = e^{r/r_o}$

natural units: $r = r_o = t_o \equiv 1 \Rightarrow 1 + z(r) = e$

proper distance: $\mathcal{R}(T) \equiv 2\,tanh\frac{1}{2}r(t)$

***World Map**: an invisible hyperboloid of co-existing objects*

$$dT^2 = dt^2 - ds^2 \ . \ \ ds^2 = dr^2 + sinh\,r^2(d\theta^2 + sin^2\theta d\phi^2)$$

The hyperbolic space of *World Map* is isotropic and homogeneous.

***World View**: a visible pseudo-sphere of shells of varying age*

$$c^2 dt^2 = dT^2 + ds^2 \ . \ \ ds^2 = \{d\mathcal{R}^2 + \mathcal{R}^2(d\theta^2 + sin^2\theta\,d\phi^2)\}/(1 - \tfrac{\mathcal{R}^2}{4})^2$$

The flat space of *World View* is isotropic, but not homogeneous; this explains the observed crowding of objects with distance:

www.astro.ucla.edu/~wright/stdystat/htm, fig.2_{1-3}.

Model M_1: "Steady State"

$$\rho \equiv sinh\,r/e^t \equiv \mathcal{R}/e^T$$

$$e^t d\rho = cosh\,r\,dr - sinh\,r\,dt = dr - sinh\,r\,dT$$

$$T = ln\{e^t/cosh^2\tfrac{r}{2}\} = ln\{e^t(1 - \tfrac{\mathcal{R}^2}{4})\} = invar.$$

$$v \equiv dr/dt \underset{d\rho = 0}{=} tanh\,r . \Rightarrow . \gamma_v \equiv 1/\sqrt{1 - v^2} \underset{d\rho = 0}{=} cosh\,r = \tfrac{dt}{dT}$$

$$\mathcal{H}_1(T) \equiv \dot{\mathcal{R}}(T)/\mathcal{R}(T) \propto constant$$

Model M_2: "Fierce Blow"

$$\rho \equiv sinh\,r/sinh\,t \equiv \mathcal{R}/sinh\,T$$

$$sinh\,t\,d\rho = cosh\,r\,dr - sinh\,r\,coth\,t\,dt = dr - sinh\,r\,coth\,T\,dT$$

$$T = arsh\{sinh\,t/cosh^2\tfrac{r}{2}\} = arsh\{sinh\,t(1 - \tfrac{\mathcal{R}^2}{4})\} = invar.$$

$$v \equiv dr/dt \underset{d\rho = 0}{=} \sqrt{sinh^2 r + \rho^2}/cosh\,r . \Rightarrow . \gamma_v \equiv 1/\sqrt{1 - v^2} \underset{d\rho = 0}{=} cosh\,r/\sqrt{1 - \rho^2}$$

$$\mathcal{H}_2(T) \equiv \dot{\mathcal{R}}(T)/\mathcal{R}(T) \propto coth\,T \underset{T \to \infty}{\to} \mathcal{H}_1$$

Model M_3: "Gentle Flow"

$$\rho \equiv sinh\,r/cosh\,t \equiv \mathcal{R}/cosh\,T$$

$$cosh\,t\,d\rho = cosh\,r\,dr - sinh\,r\,tanh\,t\,dt = dr - sinh\,r\,tanh\,T\,dT$$

$$T = arch\{cosh\,t/cosh^2\tfrac{r}{2}\} = arch\{cosh\,t(1 - \tfrac{\mathcal{R}^2}{4})\} = invar.$$

$$v \equiv dr/dt \underset{d\rho = 0}{=} \sqrt{sinh^2 r - \rho^2}/cosh\,r . \Rightarrow . \gamma_v \equiv 1/\sqrt{1 - v^2} \underset{d\rho = 0}{=} cosh\,r/\sqrt{1 + \rho^2}$$

$$\mathcal{H}_3(T) \equiv \dot{\mathcal{R}}(T)/\mathcal{R}(T) \propto tanh\,T \underset{T \to \infty}{\to} \mathcal{H}_1$$

Ungar [2008] has derived a formula obviating the need for dark matter.
Hoyle & al. [2000] explain CMBR by graphite whiskers in cosmic space.
Lerner [2020] has challenged the BB-nucleosynthesis of light elements.
BB-theory cannot explain the formation of oldest stellar structures.
For comparison, cf. Ch.5, §§ 6-7 & conclusion.

Non-Standard Relativity

ACKNOWLEDGMENT

First of all I want to thank my former student and collaborator Peter Øhrstrøm - together with whom I wrote my first published paper on relativity theory, Wegener & Øhrstrøm [1975], as well as my only published paper on tense logic, Wegener & Øhrstrøm [1996] - for a friendship that has now lasted more than four decades. Peter has written important papers of his own on relativity, e.g. Øhrstrøm [2000], before he left physics for good in favour of his passion for tense logic and its applications to AI.

In 1974, I had the good fortune to meet the renowned cosmologist and historian of science John North when he was a visiting professor at the university of my hometown Aarhus. After a conversation dr. North had the kindness to suggest my name to Julius Fraser, founder of *ISST (International Society for the Study of Time)*. In that forum I got acquainted with many other interesting and prominent people, e.g., Gerald Whitrow, Gert Müller, David Park (a dear friend), Peter Landsberg, and Paul Davies.

In the autumn of 1987, I found an advertisement in *British Journal for the Philosophy of Science*, where dr. Michael C. Duffy announced a conference on the *Physical Interpretations of Relativity Theory*, to be held at the Imperial College in London the following year. I wrote to dr. Duffy, sending him the draft of a paper, and was invited. This was to be the first of a whole series of biennial conferences, of which I attended those in London in 1988, 1990, 1992, 1994, 1996, 1998, and 2002, as well as two other conferences in Budapest in 2007 & 2009, arranged by dr. Laszlo Székely, presenting papers at them all. To participate in these conferences meant everything to me: I met numerous interesting people, of which I can only mention a few: Roger Jennison, Gope Keswani, Ludwik Kostro, Peter Kroes, Alexeí Nesteruk (my special friend), Pierre Noyes, Viv Pope, Simon Prokhovnik, David Roscoe, Peter Rowlands, Mendel Sachs, Ruggiero Santilli, Franco Selleri (a close friend), Lawrence Sklar, Maurice Surdin (a close friend), Barrie Tonkinson, Håkan Törnebohm, and Friedwardt Winterberg. In ch.1, I have praised the merits of dr. Duffy as the inspiring and indefatigable organizer of these conferences. After a few years, dr. Duffy was attacked by a severe sclerosis which made him share the tragic fate of Stephen Hawking.

In 1992, I found an article by André Mercier in *Gen.Rel.Grav.* [1975] which caught my attention, as it contained the provocative statement "*Gravitation* **is** *Time*". This induced me to write him a letter that started a long correspondence ending up with an invitation to visit him at his summer residence in France. At that time he had just lost his wife whom he had met when he studied with Niels Bohr in Copenhagen. After having visited him together with my wife in 1993, being now a member of the organizing *PIRT* committee, I seized the opportunity to invite him to participate in the 1994 *PIRT*-conference in London. Later I also succeeded in persuading him to visit us in our private home in Aarhus, in connection with a symposion I arranged in 1996 over the theme "*Time, Creation & World Order*", cf. Wegener ed. [1999]. At this special occasion he contributed a paper which he described as his "spiritual testament".

Finally, I would like to thank my friends Peter Rowlands and Abraham Ungar for their inspiring company during our stay in Budapest when participating in the second *PIRT*-conference in that city 2009. I can only regret that limited capabilities and advanced age prevent me from fully exploiting their highly ingenious contributions to contemporary mathematical physics: Rowlands [2007] & Ungar [2008].

=//=

REFERENCES
(Literature Consulted)

1. Alväger, Farley, Kjellman & Wallin [1964]:
    'Test of the 2. Postulate of SR', *Physics Lett.12*,260.
2. Aristotle (384-322): *Physics*, book iv, Loeb Class.Libr.
3. Aristotle (384-322): *De Interpretatione*, Loeb Class.Libr.
4. Arzeliés, H. [1966]: *Relativistic Kinematics*, Pergamon.
5. Augustynek, Z. [1975]: *Studia Logica 35*, 45-53.
6. Barbour, J. [2000]: *The End of Time*, Phoenix Bks.
7. Barrett, J.F. [2000]: 'Hyperbolic Geometry in Special Relativity and
    its Relation to the Cosmology of Milne', in: Duffy & Wegener, eds.:
    Recent Advances in Relativity Theory I, Hadronic Press, Fl., US, ISBN 1-57485-047-4.
8. Bell, J.S. [1986], in Davies & Brown: *The Ghost in the Atom*, Cambr.Univ.Pr.
9. Benton, M. [1959]: *The Clock Problem in Relativity* (an annotated bibliography)
    U.S. Naval Research Laboratory, Wash.D.C.
10. Bergmann, P.G. [1970]: *Found.Phys.1*, 17-22.
11. Bergson, H. [1922/1965]: *Duration & Simultaneity*, Libr.Lib.Arts.
12. Bohm, D. [1980]: *Wholeness and the Implicate Order*, Routl. & K.P.
13. Bondi, H. [1961^3]: *Cosmology*, Cambr.U.P.
14. Born, Max [1943]: *Experiment & Theory in Physics*, Dover Bks.
15. Born, Max [1950]: *Physics in my Generation*, Pergamon 1956.
16. Born, Max [1952]: *Einstein's Theory of Relativity*, Dover Bks.
17. Bridgman, P.W. [1927]: *The Logic of Modern Physics*, NY.
18. Bridgman, P.W. [1962]: *A Sophisticate's Primer of Relativity*, Wesleyan.
19. Broglie, L. de [1960]: *Non-linear Wave Mechanics*.
20. Brown & Davies [1987]: *The Ghost in the Atom*, Cambr.U.P.
21. Builder, G. [1979]: *Speculations Sci.&Techn.2*.
22. Costa de Beauregard, O. [1986]: *Found.Phys.16*, 1153.
23. Coveney & Highfield [1990]: *The Arrow of Time*, Flamingo Bks.
24. Cusanus, N. (1401-64): *De Docta Ignorantia*, Herder Verlag; cf. Wegener [1999], introd.:
    "*Unde erit machina mundi quasi habens undique centrum et nullibi circumferentia,
    quoniam eius circumferentia et centrum est Deus qui est undique et nullibi.*"
25. Davies, P.W.C. [1995]: *About Time. Einstein's Unfinished Revolution*, Penguin.
26. Davies, P.W.C. [1999], in: *Scientific American 280* no.1, Jan.
27. Dingle, H. [1964]: 'Reason and Experiment in .. STR', *Brit.Jour.Phil.Sc.15*, 41.
28. Dingle, H. [1965]: 'Relativity and Electromagnetism', *Phil.Sc.27*, 233.
29. Dingle, H. [1972]: *Science at the Crossroads*, Brian & O'Kneffe.
30. Duffy & Wegener [2000-2002]: *Recent Advances in Relativity Theory 1-2*,
    Hadronic Press, Fl., US, ISBN 1-57485-047-4 & ISBN 1-57485-050-4.
31. Duhem, P. [1962]: *The Aim & Structure of Physical Theory*, NY.
32. Eddington, A.S. [1933]: *The Expanding Universe*, Cambr. (ch.iv.2)
33. Eddington, A.S. [1935]: *New Pathways in Science*, Cambridge.
34. Eddington, A.S. [1939]: *The Philosophy of Physical Science*, Cambridge.
35. Einstein, A. [1920/1954]: *Relativity, the special & the general theory*, Methuen.
36. Einstein & Infeld [1928f.]: *The Evolution of Physics*.
37. Ellis & Bowman [1967]: Conventionality in distant sim., *Phil.Sc.34*, 16.

38. Escher, M.C. [1960]: *'Circle Limit 4'* - a most wonderful illustration of the contraction of galaxies with distance in a flat space of finite radius, cf. *www.digitalcommonwealth.org/search/commonwealth3r076t01v*
39. Essen, L. [1978]: *Wireless World,* 44-45, Oct.
40. Feyerabend, Paul [1970]: *Against Method*, New Left Pr.
41. Flood & Lockwood [1986]: *The Nature of Time*, Blackwell.
42. Fox, J.G. [1965]: 'Evidence against Emission Theories', *Amer.Jour.Phys.33*, 1.
43. Fraassen, B.v. 1985: *Intr.Phil.Time&Space*, Columb.
44. Fraser, J.T. [1981]: *The Voices of Time*, Univ.Massachus.Pr.
45. Fraser & al., eds. [1972ff.]: *The Study of Time vol.1ff.*, Springer.
46. Giedymin, J. [1982]: *Science and Convention*, Pergamon.
47. Gill & Lindesay [1993]: 'Canon. Proper Time Formul. Rel. Particle Dynamics'. *Int.J.Th.Phys.32*, 2087f.
48. Gordon, C.N. 1962: *Proc.Phys.Soc.80*, 569-592.
49. Greenaway, F. [1979]: *Time and the Sciences*, Unesco Publ., UN.
50. Grünbaum, A. [1960]: *Logical and Philosop hical Foundations of STR*, in: Danto & Morgenbesser: *Philosophy of Science*, NY.
51. Grünbaum, A. [1967]: 'Simultaneity by slow clock transport', *Phil.Sc.34*.
52. Hafele & Keating, [1972]: *Science 177*, 166-70.
53. Harrison, E.R. [1981]: *Cosmology*, Cambridge Univ.Pr.
54. Hatch, R. [1992]: *Escape from Einstein*, Kneat Company.
55. Hawking & Penrose [1996]: *The Nature of Space & Time*, Princeton U.P.
56. Hesse, M.B. [1954]: *Science and the Human Imagination*, SCM Pr..
57. Hesse, M.B. [1965]: *Forces and Fields*, Littlefield.
58. Hoyle, Burbidge & Narlikar [2000]: *A Different Approach to Cosmology*, Cambridge U.P. Apart from a presentation of the authors' own "alternative steady state theory", which I do not accept - but I agree that the observed periodic anomalies in redshifts from quasars should not just be ignored - the book marshals a series of strong arguments against the speculations supporting the Big Bang idea.
59. Hughes & Cresswell [1968]: *An Introduction to Modal Logic*, Methuen.
60. Khan, I. [1968]: 'The Principles of Reciprocity ..', *Il Nuovo Cimento LVIIB N2*, 321.
61. Kant, I. (1724-1804) [1781/87]: *Kritik der reinen Vernuft*.
62. Kant, I. (1724-1804) [1783]: *Prolegomena zu einer jeden künftigen Metaphysik.*
63. Kilmister & Tupper [1962]: *Eddington's Statistical Theory*, Oxford.
64. Kilmister, C.W. [1966]: *Sir Arthur Eddington*, Pergamon.
65. Kennicutt Jr., R.C. [1996]: *Nature 381*, 555.
66. Keswani, G.H. [1964/65], in: *Brit.Jour.Phil.Sc.15*, 286 & *16*, 21&273.
67. Kragh, H. [1996]: *Cosmology & Controversy*, Oxford.
68. Krauss, L.M. [2012]: *A Universe from Nothing*, Simon & Schuster.
69. Landsberg & Evans [1979]: *Mathematical Cosmology*, Oxf.
70. Lerner, E.J. [1991]: *The Big Bang never happened*, NY.
71. Lerner, E.J. [2020]: *235th meeting of Amer.Astron.Soc.*, cf. *spacedaily.com* 09.01.2020.
72. Lorentz, Einstein & al. [1913]: *The Principle of Relativity*, Dover bks.
73. Lorentz, H.A. [1921]: Deux mémoires de H. Poincaré, in *Acta Mathematica 38*, 293-308. An interesting evaluation of Poincaré's papers about Relativity & Qunantum Theory.
74. Lucas, J.R. [1973]: *A Treatise of Time & Space*, UP.
75. Lucas, J.R. [1999], in: J. Butterfield: *The Arguments of Time* (Cent.Vol.Brit.Acad. Oxf.UP.; to be downloaded from *http://users.ox.ac.uk/~jrlucas/* (II 26).
76. Lucas, J.R. [2002]: 'The Godelian argument', *The Truth Journal.*
77. Machamer & Turnbull eds. [1977]: *Motion & Time, Space & Matter* ..
78. Malament, D. [1977]: 'Causal Theories of Time ...', *Noûs 11*, 293.

79. Mansouri & Sexl [1977]: *Gen.Rel.Grav.8*, 497-524.
80. Marder, L. [1971]: *Time & Spacetraveller*, Penns.
81. Martin, Adolphe [1994]: *Apeiron 18*, 19.
82. McCrea & Milne [1934]: *Quart.Jour.Math.5*, 73.
83. McCrea, W.H. [1953]: 'Cosmology', *Rep.Progr.Phys.16*, 321-363.
84. Mercier, A. [1975]: Gravitation *is* Time, *General Relativity & Gravitation 6*.
85. Mercier, Treder & Yourgrau [1979]: *On General Relativity*, Akad.Verl.Berlin; cf. *www.amazon.de*.
     Can be read as a catalogue over the many serious shortcomings of Einstein's GRG
     - cf. p.149: *gravitation is not an interaction, it is time itself.*
86. Mercier, A. [2000]: 'The Reconstruction of Spacetime as Timespace', in:
     Duffy & Wegener: *Recent Advances in Relativity Theory vol.I*, Hadronic Press, Fl. US.
87. Merleau-Ponty, J. [1965]: *Cosmologie du XXme siecle*, Gallimard;
     describes Milne's *KR* as a Leibnizian monadology put into mathematics.
88. Milne, E.A. [1935]: *Relativity, Gravitation & World-Structure*, Oxf.Univ.Pr.
89. Milne, E.A. [1948/1951]: *Kinematic Relativity*, Oxford Univ.Pr.
90. Milne & Whitrow: [1949]: On the so-called "Clock-paradox", *Phil.Mag.40*, 1244.
91. Milne, E.A. [1949]: *Gravitation without Gen.Rel.*, in P.A. Schilpp, ed.:
     Alb. Einstein, Philosopher-Scientist, Library of Living Philosophers Vol.7.
92. Milne, E.A. [1950]: 'The relativity of Galilean frames', *Proc.Roy.Soc.200*, 219.
93. Milne, E.A. [1952]: *Modern Cosmology and the Christian Idea of God*, Oxf.
94. Misner-Thorne-Wheeler [1970]: *Gravitation*, Freeman.
95. Mittelstädt, P. [1975]: *Zeitbegr.i.d.Rel.Theor.*
96. Mohr, Georg [1672]: *Euclides Danicus*, Amsterdam (german transl. Copenhagen 1928) -
     anticipated the insights of L. Mascheroni [1797]: *Geometria del compasso*, by 125 years!
97. Moon & Spencer [1953]: 'Binary stars and the velocity of light', *Jour.Opt.Soc.Am.43*, 635.
98. Moon & Spencer [1956]: 'On the establishment of a universal time', *Phil.Sc.23*.
99. Møller, C. [1972]: *The Theory of Relativity*, Oxford.
100. Narlikar, J.V. [1980]: 'Non-Standard Cosmologies', *Fund.Cosm.Phys. 6*, 1-186.
101. Narlikar, J.V. & Burbidge, G. [2008]; *Facts and Speculations in Cosmology,* Cambridge U.P.
     Offers a more recent and popular version of the arguments in Hoyle, Narlikar & Burbidge [2000].
102. Nordenson, H. [1969]: *Relativity, Time & Reality*, Ld.
103. Norman & Setterfield [1987]: *Atomic Constants*, Stanf.
104. North, J.D. [1965]: *The Measure of the Universe*, Oxf.Univ.Pr.
     One of the best surveys of the basic ideas and theories in modern cosmology.
105. North, J.D. [1970]: 'The Time-coordinate in Einstein's Restr.Th.Rel. *Studium Generale 23*, 203.
106. Novikov, I. [1980]: *The River of Time*, Canto Bks., Cambr.UP.
107. Osborne, A..D. & Pope, N.V. [2007]: *Light Speed, Gravitation & Quantum Instantaneity*,
     Philosophical Enterprises, Swansea, UK. The idea of quantum instantaneity is important.
108. Øhrstrøm, P. [1980]: *Found.Phys.10*, 333-43.
109. Øhrstrøm, P. [1984]: *Erkenntnis 21*, 209-22.
110. Øhrstrøm & Hasle [1995]: *Temporal Logic*, Kluwer.
111. Øhrstrøm, P. [2000]: 'Tense Logic and Special Relativity', in:
     Duffy & Wegener: *Recent Advances in Relativity Theory vol.I*, Hadronic Press, Fl. US.
112. Palacios. J. [2006]: *Antirelativistic Dynamics & Relativity: A Paradox-Free Variant*,
     posthumous edition by A.F. Kracklauer: Lulu.com, Morrisville, NC.
113. Pearson & al. [1981]: *Nature 290*, 365.
114. Penrose, R. [1989]: *The Emperor's New Mind*, Oxford U.P.
115. Phipps, T.E. [1986]: *Heretical Verities* .., Class.Non-Fict.Libr., Urbana, Ill.
     A furious, but well argued, attack on the dogmatism of the scientific establishment.

Non-Standard Relativity

116. Phipps, T.E. [1994]: *Galilean Electrodyn.5*, 46.
117. Phipps, T.E. [2006]: *Old Physics for New*, Apeiron.
118. Pierseaux, Y. [1999]: *La "structure fine" de la Relativité Restreinte*, L'Harmattan.
119. Pierseaux, Y. [2009]: Paper contributed to PIRT Budapest 2009.
120. Plato (427-347): *Parmenides*, Loeb Class.Libr.
121. Plato (427-347): *Timaios,* Loeb Class.Libr.
122. Poincaré, H. [1952]: *Science and Hypothesis*, Dover Bks.
123. Poincaré, H. [1958]: *The Value of Science*, Dover Bks.
124. Poincaré, H. [1952]: *Science and Method*, Dover Bks.
125. Poincaré, H. [1963]: *Mathematics and Science*, Dover Bks.
126. Poincaré, H. [1905]: 'Sur la Dynamique de l'Electron', *Comptes Rendus 140*, June 5th,1504-08.
    In fact, Poincaré's *Relativité Restraimte* anticipated Einstein's *SR* by more than three weeks.
127. Poincaré, H. [1906]: 'Sur la Dyn. de l'Electr.', *Rendiconti del Circ.mat.d.Palermo 21*,129-76.
128. Poincaré, H. [1908]: *Revue générale des Sciences pures & appl. 19*, 386.
129. Pope, V. [1994], in: *Proceedings, Physical Interpretations of Relativity Theory*,
    British Society for the Philosophy of Science.
130. Pope, V. [1996]: 'The tantalizing two.slit experiment', in:
    Duffy & Wegener, eds. 2002: *Recent Advances in Relativity Theory 2,*
131. Popper, K.R. [1958]: *The Logic of Scientific Discovery*, Ld.
132. Popper, K.R. [1963]: *Conjectures and Refutations*, Routledge.
133. Popper, K.R. [1972]: *Objective Knowledge*, London.
134. Popper, K.R. [1982]: *Quantum Theory and the Schism in Physics*, London.
135. Prigogine, I. [1983]: *From Being to Becoming*, Freeman.
136. Prior, A.N., [1962]: *Formal Logic*, Oxford - his argument p.70 shows that Rowlands cannot escape Gödel:
    We cannot prove the statement which results from replacing the variable in the statement.form:
    'We cannot prove the statement which results from replacing the variable in the statement.form y by the name
    of the statement-form in question' by the name of the statement-form in question. - cf. this book Ch.8, Q.12.
137. Prior, A.N., [1967]: *Past, Present & Future*, Oxford U.P.
138. Prior, A.N., [2003]: *Papers on Time & Tense* (Hasle, Øhrstrøm, Braüner, Copeland, eds.), Oxford U.P.
139. Prokhovnik, S.J. [1967]: *The Logic of Special Relativity*, Cambridge U.P.
    A very informative introduction to the intricacies of the special relativity theory.
140. Prokhovnik, S.J. [1973]: *Found.Phys.3*, 351.
141. Prokhovnik, S.J. [1974]: *Int.Jour.Th.Phys.9*,291.
142. Prokhovnik, S.J. [1976]: *Found.Phys.6*, 687-705.
143. Prokhovnik, S.J. [1980]: *Found.Phys.10*, 197-208.
144. Prokhovnik, S.J. [1988]: 'Nature & Implications of **RWM**', in: Duffy & Wegener, eds.:
    Recent Advances in Relativity Theory I, Hadronic Press, Fl., US, ISBN 1-57485-047-4.
145. Quine, W.v.O. [1953]: *From a Logical Point of View*, Harvard.
146. Rae, A. [1986]: *Quantum Physics: Illusion or Reality?* Cambridge U.P.
147. Ramakrishnan, A. [1973]: *J.Math.Anal.42*, 377.
148. Reichenbach, H. [1924]: *Axiomatik der relativ. Raum-Zeit Lehre*.
149. Reichenbach, H. [1927/1958]: *The Philosophy of Space & Time*.
150. Rescher, N. [1967]: *The Philosophy of Leibniz,* Pr-H.
151. Rietdijk, C.W. [1960]: 'A Rigorous Proof of Determinism ..', *Phil.Sc.33*, 341.
152. Rindler, W. [1960]: *Special Relativity*, Oliver & Boyd.
153. Rindler, W. [1968]: *Am.Jour.Phys.36*, 540.
154. Robertson, H.P. [1935&36]: *Astrophys.Jour.82&83*, 187&257.
155. Romain, J.E. [1963]: *Il Nuovo Cim.30*, 1254.
156. Rowlands, P. [1994]: *A Revolution too far. The Establishment of **GR***, PD Publications.

157. Rowlands, P. [2007]: *Zero to Infinity. The Foundations of Physics*, World Scientific.
158. Rowlands, P. & Hill, V. [2012]: *Fund. math. structures applied to physics and biology* (search Google)
159. Roxburgh & Tavakol [1975]: *M.N.Roy.Ast.Soc.170*,599.
160. Russell, B. [1925]: *The ABC of Relativity*, Allen & Unwin.
161. Sachs, M. [1993]: *Relativity in Our Time*, Taylor & Francis.
162. Schiff, L.I. [1960]: *Am.Jour.Phys.28*,340.
163. Schutz, J.W. [1973]: *Foundations of Special Relativity*, Springer (Lect.Notes Math.361).
164. Selleri, F., see *http://www.ba.infn.it/~selleri/*
165. Selleri, F. [1997], in: *Found.Phys.Lett.10*, 1.
166. Selleri, F. [2009]: *Weak Relativity* (forthcoming)
167. Shapiro, I.I. [1972]: *Gen.Rel.Grav.3*, 135.
168. Sherwin, C.W. [1960]: *Phys.Rev.120*, 17.
169. Sitter, W. de [1913]: 'A proof of the const.of vel.light', *Proc.Acad.Amsterd.15*, 1297 & *16*, 395.
170. Sklar, L. [1985]: *Philosophy & SpaceTime Physics*, Calif.
171. Smith, S.L., author of the brilliant math text processor **EXP** (only compatible with **Win XP** or lower)
172. Smolin, Lee [1997]: *The Life of the Cosmos*, Oxford U.P.
173. Smoot & al., [1977]: *Phys.Rev.Lett.39*, 898f.
174. Stephenson & Kilmister [1958]: *Special Relativity for Physicists*, Dover Bks.
175. Strauss, M. [1966]: *Wiss.Zeitschr.Univ.Jena 15*, 109.
176. Stump, D. [1989], in: *Stud.Hist.Phil.Sc.20*, 335-363.
177. Surdin, M. [1978]: *Found.Phys.8*, 341.
178. Surdin, M. [2002]: 'Cosmology & Stochastic Electrodynamics', in:
    Duffy & Wegener, eds.: *Recent Advances in Relativity Theory 2,* Hadronic Press, Fl., US.
179. Synge, J.L. [1960]: *Relativity, General Theory*, Amst.
180. Tonkinson, B.J. [2000]: 'Clocks don't go slow, rods don't contract', in:
    Duffy & Wegener, eds.: *Recent Advances in Relativity Theory 1,* Hadronic Press, Fl., US.
181. Törnebohm, H. [1957]: 'Epistemol. Reflections on Milne's .. Two Times', *Phil.Sc.24*,57.
182. Törnebohm, H. [1963-4 & 1971], in: *Theoria 26*, 79,283,417 & *Theoria 37*, 209
183. Törnebohm, H. [1974]: *Minkowskian Spacetime Th.*, Univ.Göteborg.
184. Törnebohm, H. [1963]: *Concepts & Principles .. within SR*, Acta Univ. Gothob., Sthlm.
185. Törnebohm, H. [2000]: 'Infra-theories to the special theory of relativity', in:
    Duffy & Wegener, eds.: *Recent Advances in Relativity Theory 1,* Hadronic Press, Fl., US.
186. Trempe, Jacques [1990]: *Apeiron 8*, 1.
187. Trempe, Jacques [1992]: *Phys.Ess.5*, 1.
188. Ungar, A.A. [1986]: *Phil.Sc.53*, 395.
189. Ungar, A.A. [2008]: *Analytic Hyperbolic Geometry and A.E.'s STR*, World Scientific.
190. Van Fraassen, B. [1989]: *Laws and Symmetries*, Oxford.
191. Varičak, V. [1924/2007]: *Relativity in three dimensional Lobachevsky space*,
    Lulu.com, Morrsiville, NC, ISBN 978 1 84753 364 7.
    Varičak, whose work inspired Ungar, argued that the natural 3-space of SR must be hyperbolic.
192. Voigt, W. [1887]: *Nachr.Kön.Ges.Wiss.Göttingen 2*, 41.
193. Waddoups, Edwards & Merrild [1965]: 'Exper.Invest. of 2.Post.STR, *J.Opt.Soc.Amer.55,* 142.
194. Walker, A.G. [1935]: *Monthl.Not.Roy.Astr.Soc.95*, 263.
195. Walker, A.G. [1936]: *Proc.Lond.Math.Soc.42*, 90.
196. Walker, A.G. [1940 & 1943]: *Proc.Lond.Math.Soc. 46*, 113-154 & *48*, 161-179.
197. Walker, A.G. [1944]: *Proc.Lond.Math.Soc. 19*, 219-29.
198. Walker, A.G. [1948]: *Proc.Roy.Soc.Edinb.62A*, 319-335.
199. Walker, A.G. [1959]: 'Axioms for Cosmology', in:
    Henkin, Suppes & Tarski: *The Axiomatic Method*, North Holland.

200. Wallace, B.G. [1971]: 'Radar-evidence that the velocity of light .. is not c', *Spectrosc.Lett.4*, 79.
201. Wegener & Øhrstrøm [1975]: *Il nuovo cimento 30B*, 291.
202. Wegener, M.T. [1990]: Poincaré's Theory of Restricted Relativity, in: *PIRT-proc.*, BSPS.
203. Wegener, M.T. [1993]: 'Relativity with Absolute Simultaneity', *PIRT-proc.add.*, BSPS.
204. Wegener, M.T. [1993]: 'Time and Harmony in Leibniz', in G. Seel & al., eds.:
      L'Art. la Science et la Metaphysique (Festschrift to André Mercier), Peter Lang.
205. Wegener, M.T. [1994]: *PIRT-Proceedings*, BSPS, London, late papers
206. Wegener, M.T. [1994]: *PIRT-Proceedings*, BSPS, London, supplement (a,b,c,d)
207. Wegener, M.T. [1994]: *Philosophia Scientia I, cahiers spécial 1* (ACERHP, Nancy, Fr.).
208. Wegener, M.T. [1995a]: 'The Radar Technique as a Theoretical Device', *Physics Essays 8*.
209. Wegener, M.T. [1995b]: 'A Classical Alternative to Special Relativity', *Physics Essays 8*.
      The theory in [1995a&b] is interesting in its own right, but I no longer consider it to be viable;
      the basic ideas are similar to, but independent of, those of Trempe [1990f] and Martin [1994].
      Tepper Gill, who has worked on proper time formulations of classical & relativistic ED,
      has kindly acknowledged inspiration from an earlier version (Google TG, *howard.edu*).
210. Wegener & Øhrstrøm [1996]: 'A New Tempo-Modal Logic for Emerging Truth',
      in: Faye & al.: *Perspectives on Time*, Kluwer (Boston Studies in Phil.Sc.,189).
      For a fuller version of the system ***W***, cf. ***www.M-T-W.me***, 'A Logic of Creation'.
211. Wegener, M.T. [1999], ed.: *Time, Creation & World-Order*, Aarhus Univ.Pr.
212. Wegener, M.T. [2000]: 'Ideas of Cosmology: A Philosopher's Synthesis', in:
      Duffy & Wegener, eds. 2000-02: *Recent Advances in Relativity Theory 1,* Hadronic Press, Fl., US.
213. Wegener, M.T. [2002]: 'Some Cosmological Models: Their Time-Scales & Space-Metrics', in:
      Proceedings PIRT 8, Imperial College, London.
214. Wegener, M.T. [2004]: 'The Idea of a Cosmic Time', *Found.Phys.34*, 1777-99.
215. Wegener, M.T. [2006]: 'Kinematic Cosmology', in: *1st Crisis in Cosmology Conf.*, AIP 822, NY
      (regrettably, this paper is marred by some formal opacity).
216. Wegener, M.T. [2007]: 'Big Bang versus Steady State', PIRT, Budapest.
217. Wegener, M.T. [2009]: 'New Axioms for Cosmology', PIRT, Budapest.
218. Weizsäcker, C.F.v. [1985]: *Aufbau der Physik*, Carl Hanser.
219. Weizsäcker, C.F.v. [1992]: *Zeit und Wissen*, Carl Hanser.
      His problematic behaviour under WW2, of course, does not invalidate his scientific views.
220. Wesley, J.P. [1980]: *Found.Phys.10*, 503-510.
221. Wesley, J.P. [1986]: *Found.Phys.16*, 817-824.
222. Whitrow, G.J. [1946]: *Philosophy 21*.
223. Whitrow, G.J. [1952]: *Sc.Proc.Roy.Dubl.Soc.26,1,* 37-44.
224. Whitrow, G.J. [1959]: *The Structure and Evolution of the Universe*, Harper NY.
225. Whitrow, G.J. [1965]: *Vist.Astron.6*, 1.
226. Whitrow, G.J. [1961/1980]: *The Natural Philosophy of Time*, Nelson/Oxford.
227. Whitrow, G.J. [1972]: *What is time*, Thames & Hudson (Penguin?).
228. Whitrow, G.J. [1973]: Inaugural Lecture, Univ.Ld.
229. Winnie, J. [1970]: 'SR without one-way assumptions', *Ph.Sc.37*, 81&223.
230. Whittaker, E.T. [1947]: *From Euclid to Eddington*, Harper; Dover bks.1958.
231. Whittaker, E.T. [1953]: *A History of the Theories of Aether & Electr.II*, Harper.
232. Yolton, J.W. [1960]: *The Philosophy of Science of A.S. Eddington*, Nijhoff.

=//=

INDEX

=//=

WISE & WITTY
words by my friend

TOM PHIPPS, ARCH-HERETIC
on relativity theory

"*You don't get physical garbage out of any mathematical theory without putting it in at the start. Actually the 'physics' deduced in such cases is invariably a form of emergency surgery to stop arterial bleeding of the logic of the theory. According to Einstein, GRT was solidly built upon SRT. SRT was built upon c as a limiting speed in nature. And GRT, without contradicting SRT, 'predicts' - in flat contradiction of SRT - that something called 'space' long ago exhibited a physical property of spectacular inflational elasticity but, in agreement with SRT, no longer does so today because, if it did, we would measure speeds greater than c in our lab. However, we can look at galaxies in opposite directions today and see this elasticity at work - while, according to SRT's 'worldline' concept, long ago and today and the distant future are all the same, any distinction being physically meaningless (because observers in different states of motion disagree about them). And if long ago and today are the same ..., because of spacetime symmetry, so separations of objects in our lab and of distant galaxies are the same - and lab space is elastic, after all, like the critical sense of the relativist. If you buy all or any of that, there is a bridge in Brooklyn I'd like to sell you ... and a tonic that long ago would have grown hair on a billiard ball, though not today - except that long ago and today are the same, so it might be worth an open-minded trial at your risk, $179.95 the bottle plus postage!*"

T.E. Phipps jr.**,** 2006: ***Old Physics for New***, p.219.